The Edge

Jonathan Maxwell

The Edge

How competition for resources is pushing the world, and its climate, to the brink – and what we can do about it.

Jonathan Maxwell

First published by Nicholas Brealey Publishing in 2023
An imprint of John Murray Press

1

A CIP catalogue record for this title is available from the British Library

Hardback ISBN 978 1 39981 0 845
ebook ISBN 978 1 39981 0 869

Typeset by KnowledgeWorks Global Ltd.

Printed and bound in Great Britain by Clays Ltd, Elcograf S.p.A.

John Murray Press policy is to use papers that are natural, renewable and recyclable products and made from wood grown in sustainable forests. The logging and manufacturing processes are expected to conform to the environmental regulations of the country of origin.

John Murray Press
Carmelite House
50 Victoria Embankment
London EC4Y 0DZ

Nicholas Brealey Publishing
Hachette Book Group
Market Place, Center 53, State Street
Boston, MA 02109, USA

www.nicholasbrealey.com

John Murray Press, part of Hodder & Stoughton Limited
An Hachette UK company

To my wife, Laura, and my children, Zachary and Charlie.

Like everything else in my life, this book would not exist
without Laura.

Contents

Prologue

Why are our energy and food bills skyrocketing?

Why is there a cost-of-living crisis?

Why is inflation so high, devaluing our hard-earned savings?

Why might the lights go out?

Why might there not be enough energy to go around?

Why are there water shortages?

Why do people die from air pollution in London and New York?

Why are our jobs at risk? Are our children safe?

Why do people die of cold in the winter?

Why are the rivers and lakes running dry?

Why is it so hot in the summer? Why are there so many fires?

Why was the family member of our Ukrainian friends killed?

Why are there so many migrants and refugees?

Why are so many people worried about climate change?

Why is what we are doing making things worse?

Why can't renewables and vegetables save the world?

Why are we wasting most of the energy and food in the world?

Why is now a turning point in history?

Why are we, quite literally, on the edge?

These things matter to us. I am going to try to explain 'why'.

And I am also going to try to explain 'so what'.

Well-meaning policies, which often come from deeply held morality-based convictions, can sometimes make the problems they are trying to solve considerably worse. There is a danger that responses to the climate crisis that place reliance on, for instance, adding new sources of energy to the system over unrealistic time-frames and without replacing or reducing conventional sources risk losing time and money that we cannot afford to waste.

There are connections and parallels to draw between the climate crisis and geopolitical crises. Demand and competition for energy and resources are both drivers of environmental and climate crises and, at the same time, causes of and flames for geopolitical crises and events. Russia's invasion of Ukraine in 2022 was a turning point, indeed a point of no return. At the heart of the energy system, the regional conflict created the world's first global energy crisis.

So, what do we do about it? I will make the case for resource efficiency as the top priority for all of us and a lens through which we should place actions that governments, companies and individuals can take in order of merit. Applying this lens to look at environmental, climate, societal and political problems in this way, each of us can influence our politicians, corporate management and our friends and family to make choices that make a positive difference. We can take some of society's greatest challenges and create some of its greatest opportunities. We can find in resource efficiency the ingredients for extraordinary productivity and growth and, at the same time, a healthier and safer future.

Then, we must be practical, realistic, fast, creative and ambitious. And we must always seek to understand the limits and act within them.

Introduction

A definition of sustainable development: 'Development that meets the needs of the present without compromising the ability of future generations to meet their own needs.'

Our Common Future *or the* Bruntland Report, *published in 1987 by United Nations' World Commission on Environment and Development.*

We are at the edge, not the end, of history.

The world is reaching its limits and is at war over the critical resources that define them. Fighting over energy has now spilled over the edge of Europe but rages quietly, but with ever increasing intensity, between great financial powers over coveted resources. The economy, everyone and everything depends on energy, as well as the minerals and water needed to generate it. Everyone depends on food, which is being cut off by conflict over it.

The Edge is about our limits, but also a turning point. Have we reached one now, why, so what, and what should we do about it?

Russia's invasion of Ukraine was a watershed of a turning point. Ukraine, which means literally 'on the edge', had been a hidden arena of competition for resources by nation states and energy markets and was now transformed into a coliseum. Stunned Europeans recoiled at the aggression on their doorstep, while war evoked the trauma that Europe had sworn that it would endure 'never again'. But beyond the battlefield, this was a resource war, the manifestation of great geopolitical shifts in the balances of power between East and West.

The war sent shockwaves through the power system, both political and physical. It was not just the supply lines to the battlefield that were at stake, it was the battlefield that was a supply line to the rest of the intimately interconnected world. Europe's energy and the Middle East and Africa's food were on the military planning table, while Russia, China and the USA pored over it. To physical security were added the problems of energy security and food security. And then, it was revealed, the supply of the minerals and other resources which only a few ever thought much about, but on which the whole world had begun to depend. As everything had changed, so did geopolitics and so did global markets. Prices exploded.

Politics, say historians, is power. Our politics today, in an intimately interconnected world, is geopolitics. Power, literally, comes from energy, on which everyone and everything depends. While asking how to replace the supplies of Russian natural gas that the world relied on before the invasion of Ukraine, we must start to ask other questions, such as why are we wasting most of the world's energy and why don't we do something about it? Not just to save money and carbon and improve resilience, but to reduce the risk, or effect, of conflict.

Competing for and consuming energy resources beyond limits also creates threats to life as more carbon is generated than the planet can absorb, changing the climate. But hopes of defying one set of limits associated with conventional energy by turning to seemingly limitless renewable energy are being frustrated. Neither the time nor the resources needed to make renewable energy are unlimited. We will run out of carbon budget by the end of the decade.

While all eyes are on making more of everything, generating, replacing, and adding to meet fast-growing and competing needs, the limits of time and resources show us that we simply cannot meet these needs on the supply side alone. By seeking solutions through adding more, we are looking the wrong way. For all our efforts to create more, we miss the basic fact that we waste most of the scarce energy, food and water resources that we use and compete for.

We are heading in the wrong direction, fast, and accelerating towards the edge, the limits. As we cross boundaries on the physical and financial battlefields and with the climate and our environment, we are now moving into uncharted territory, taking big chances, and stacking the odds of surviving the journey against ourselves. But if we act fast, change direction and focus on reducing consumption by reducing waste, not adding to it, choosing efficiency and productivity, then we stand the best possible chance of changing the odds back in our favour.

I trained as a modern historian at Oxford University. The very last lesson I learned was possibly the most important. The only question that you really had to ask, they said, is 'so what?' I have carried this question with me ever since, together with the associated question: 'why?' I now ask this question every day as a professional investor. After building a career in finance, I left my job as an investment director at HSBC to follow my convictions. I put everything on the line to set up an investment firm, Sustainable Development Capital LLP, or 'SDCL', in 2007. This was before 'sustainability' was a 'thing' or even a talking point, but it was an important year, a turning point for the world, in many ways perhaps imperceptibly. But setting up my own firm enabled me to focus on key questions and ask, for example, why we waste most of the world's energy. Eventually, I was able to secure investment, now billions, to do something about it. *The Edge* is therefore, in many respects, a personal story. It draws on the past 15 years of my life at the heart of the energy and climate change sector. It looks inside the mindsets of government, big business and finance, and suggests what it takes to change their minds.

The question 'why?' requires us to look behind problems and ask difficult questions. In looking behind the change in the world order today, we gravitate to the confluence of two key determining factors: economics and the environment, defining ingredients of modern geopolitics. The 'so what?' is to establish a framework, an approach to making decisions and taking actions, as modern societies, towards, as the likes of [polymath] academic and author of *Collapse* (2006) Jared Diamond suggested, choosing to survive rather than fail when confronted with these limits. At least our generation now has the benefit of hindsight, as well as the technology, to recognize and avoid societal collapse from environmental changes, the effects of climate change, competition and conflict.

The period during which human activities have had an environmental impact on the Earth is now regarded as a distinct geological age, the Anthropocene, often traced back to the Industrial Revolution. The Industrial Revolution was, is, characterized by innovation, agriculture and exploitation of natural resources, for instance in the generation of energy for machines. Energy powered the Industrial Revolution, but we have now reached an edge, an outside limit for our resources, and at the same time a turning point both for geopolitics and for the ecosystem that sustains our planet.

Scientists have now identified nine 'planetary boundaries' to define a framework for these limits. Crossing a planetary boundary comes at the risk of abrupt environmental change. The framework is based on scientific evidence that human actions, especially those of industrialized societies since the Industrial Revolution, have become the main driver of global environmental change. Under the framework, 'transgressing one or more planetary boundaries may be deleterious or even catastrophic due to the risk of crossing thresholds that will trigger non-linear, abrupt environmental change within continental-scale to planetary-scale systems'.[1] In 2022, scientists considered that we had crossed five of the nine: climate change, loss of biosphere integrity, land-system change, altered biogeochemical cycles (phosphorus and nitrogen) and the introduction of novel entities. Measures of our 'ecological footprint' suggest that our level of consumption outweighs our resources by 1.75:1.

Most often, sustainable development is discussed in terms of energy, with a focus on identifiable human-induced climate change. This is because it is now understood that the 'carbon cycle' is out of balance because more carbon (shorthand for greenhouse gas) is being emitted than can be absorbed. The accumulation of gases, including water vapour, nitrous oxide, methane, and carbon dioxide (CO_2), creates a global 'greenhouse' around the planet, trapping more heat from the sun than otherwise flows back into space, warming the Earth. Concerns that the cycle is perpetuating, increasing the level of gas that does not dissipate for years, have led to a consensus of fear of negative feedback loops, 'runaway' climate change and catastrophic effects on severe weather, land, oceans and sea levels.

Hence the 'energy transition', a massive movement to reduce fossil fuel consumption in favour of cleaner and renewable energy, with the objective of limiting global temperature rise to 1.5°C from pre-industrial levels by reducing human carbon emissions to net zero. However, so far, the global energy transition has actually been the 'phase of energy addition', as a leading energy strategist put it.[2] Previous energy transitions involved the shift from wood and crop residues to coal, then from coal to oil by the 1960s, then from oil to gas in the 2000s. After falling for most of the twentieth century, with a reducing share of bioenergy, renewable energy production, particularly from wind and solar, has been accelerating recently in the last 20 years, but this is in addition to growing production of conventional energy. Renewable energy has a lower marginal carbon footprint than most conventional generation, but it is not zero and involves the use of other limited resources, such as rare earth metals. This can lead to scarcity, competition for resources and, like the competition for energy, even conflict.

Where resources are limited, this demands efficiency. The overwhelming majority of all effort from countries and companies is focused on the supply side. But we must now also focus as much on the demand side of the equation, reducing the size of the cause, not just the symptoms. When we understand the limits of supply of critical natural resources and the reasons for them – 'why?' – including systemic waste, we can draw conclusions, consider the implications – 'so what?' – and conclude that efficiency is key. Breaching the limits of the world's natural resources creates unsustainable environmental damage and geopolitical conflict. We must now step back from these limits and focus on being more efficient with how we use resources. This is the only pathway for sustainable growth on the one hand and reduction of conflict on the other.

The Edge seeks to weave together what might otherwise seem like unconnected issues and stories into a pattern that reveals the intricate interdependency of key elements of modern society: energy, resources, climate, the environment, human, natural and financial capital, technology, progress and, perhaps as importantly for our own interests, the way we interpret and tell the story itself: culture. As the Nobel laureate Hermann Hesse once complained, 'contemporaries are never able to see their own place in the patterns'. At this crucial turning point, where we have reached an edge, we need to do better.[3]

In many ways, the story of *The Edge* starts a little over 15 years ago, in 2007. While I was busy setting up SDCL in London, deep underwater adventures were taking place in the Arctic, as Russia was planting a titanium flag a thousand leagues below the North Pole under the watchful eye of President Vladimir Putin. Some 10 years later, with the annexation of Ukraine's Crimea under his belt, the same eyes personally oversaw the loading of a massive ice-breaker tanker carrying the first cargo of oil that Russia extracted from the Arctic, ironically circumnavigating US sanctions to arrive on a cold evening in a blue-hulled tanker that docked on the Mystic River in Boston, Massachusetts. The USA achieved energy independence the following year and then, with the COVID-19 pandemic and the Beijing Olympics all but over, Russia invaded Ukraine. Energy and commodity prices, already high, spiked, sending inflation soaring and nearly blowing out parts of the financial system, like the London Metal Exchange, which was pushed to the brink by a massive bet placed by a Chinese industrial group and bailed out by the Hong Kong-owned exchange. Meanwhile, tensions were rising in the South China Sea. Chapter 1 and Chapter 2 look at the background to these stories and the tension and intense competition related to energy and resources, demand for which has reached the very limits of supply, that create the conditions for conflict that sit behind the chapters that follow.

Chapter 3 looks at the climate, which is changing. Understanding about the climate is changing too. As scientific consensus sets in as to what is causing climate change and how humanity is contributing to it, there remains no such consensus about what to do about it. Indeed, some well-meaning reactions might well make it worse, while the flight to energy security and diversification following Russia's invasion of Ukraine threatens to set the agenda back. In the meantime, we don't have the time to waste as we race to implement change before our carbon budget burns out. Where most of the human-induced problem comes from energy, there is a case to be made that less, rather than more, fuel is needed for the fire.

Chapter 4 deals with the economics of climate change. The costs of suffering from climate change have been set in conflict with the costs of preventing it. The case is made that making the right decisions to protect the environment, resources and the climate does not necessarily mean reducing economic performance but can instead be a source of productivity and growth in itself. Chapter 5 deals with the paradox of addressing a problem associated with limited resources with a solution – renewable energy – that can, perhaps seductively, seem to be unlimited, but is certainly not. Also, given the scale of the challenge at hand, the solution will not be 'all of this' or 'all of that' but 'all of it'. Chapter 6 looks at the problem the other way round and investigates how the size of the problem can and must be reduced, through efficiency first.

Chapter 7 looks at the natural world and the natural resources on which humanity entirely depends. Not all these natural resources are renewable, and those that are naturally renewable might cease to be if they are pushed over the edge. Those that are not renewable are also the stuff that society relies on for survival, the story of the food system, which relies almost entirely on fossil fuels today, being amongst the most illustrative. Chapter 8 looks at human capital, which after all is what we are all living and fighting for. As humanity moves into town, it looks at what urbanization and economic 'progress' have in store for us. Chapter 9 takes us into the mind set of big business and finance and looks at how it thinks, rightly or wrongly, it is helping, but also how the inherent conflicts have set the stage for political, legal, and financial contests.

Chapter 10 looks at the state of technology and innovation and whether, how and when they might help advance the state of human knowledge, which may currently be slowing rather than accelerating, and deal with the enormous challenges identified in *The Edge* so far. It looks at how far we have come in the past 15 years, in some respects, and how little in others. Chapter 11 looks at how we talk about climate and the environment, how this can backfire as powerfully as policy or practice itself, with equally significant impact, and how we might be able to turn this around. Chapter 12 lays a framework for a brighter and more sustainable future, an opportunity for better growth built on productivity, prosperity and efficiency.

The epilogue tells my story, the journey so far. Fifteen years ago, at the beginning of the journey, the United Nations' Intergovernmental Panel on Climate Change and former US vice president Al Gore were demonstrating with greater certainty that humankind was behind climate change and telling the story that earned the Nobel Prize. Fifteen years later, the science has become conclusive and the voracious consumption of resources that has caused climate change has now exploded into open conflict. During that time, my journey bore witness to the enormous challenges that the world and its institutions face, but also the extraordinary opportunities that lie ahead of us.

CHAPTER 1

Energy Collides with History

• • •

'The world will not be destroyed by those who do evil, but by those who watch them without doing anything.'

Albert Einstein

- Russia's invasion of Ukraine needs to be understood in the context of long-term competition for natural resources.
- The risk of other geopolitical conflicts over critical natural resources should be expected and mitigated.
- Competition for natural resources is also a defining factor in the degree to which we succeed or fail in addressing the climate crisis.

On 2 August 2007, two Russian mini submarines, both Mir deep-submergence vehicles, Mir I and Mir II, had reached the seabed of the Arctic Ocean, more than 4 kilometres beneath the North Pole. Mir II's crew comprised a Russian pilot, an Australian adventurer, and a Swedish businessman. Mir I was piloted by Anatoly Sagalevich, a researcher at the Russian Shirshov Institute of Oceanology Institute, businessman Vladimir Gruzdev, and Artur Chilingarov, an acclaimed Russian engineer-oceanographer, polar explorer and member of the Russian State Duma. Chilingarov had been awarded the title of Hero of the Soviet Union on 14 February 1986 for his success, and display of organizational abilities and courage, in rescue operations in extreme conditions.

Chilingarov was on a mission. He was jointly leading and had helped finance this expedition for a very special purpose. Mir I dove into the dark and freezing depths of the polar Arctic waters at 09:28 and, more than two and a half hours later, reached the seabed 4,261 metres below the ice. On the seabed, Mir I planted a one-metre-high titanium Russian flag, with echoes of the USA's moon landing on 20 July 1969. The flag, which was specially designed at Kaliningrad's 'Fakel' design bureau, was an expression of achievement but it was also a claim of ownership. It was left with a time capsule containing a message for future generations and the flag of the pro-President Putin United Russia party. The flag had been planted on the Lomonosov Ridge, which Moscow claims is directly connected to its continental shelf. Russia's foreign minister, Sergei Lavrov, commented on the scene to Radio Mayak: 'The goal of this expedition is not to stake out Russia's rights, but to prove that our shelf stretches up to the North Pole ... There are concrete scientific methods for this.'[1] The United Nations Convention on the Law of the Sea defines an area 200 miles beyond a country's coast as its exclusive economic zone, or 350 miles if a country can prove a 'continental margin' – a shallow shelf linked to the mainland. America (through Alaska), Canada, Denmark (through Greenland), Finland, Iceland, Norway, Sweden and Russia all have territory within the 16,000 km Arctic Circle. (A NATO aircraft carrier was sent into the Arctic Circle for the first time in 27 years in 2018. In 2021, British and American warships entered the Barents Sea for the first time since the 1980s. Norway conducted its biggest military exercise in the Arctic since the Cold War in May to June 2023. The exercise, called Arctic Challenge Exercise 2023, involved around 150 aircraft and 3,000 personnel from 14 countries.

The ascent to the icy surface was even more difficult and dangerous than the long and deep dive to the seabed because, on the return, the Mirs had to locate a small hole on the surface as the vessels themselves were too small to break through the ice on their own. If they failed, they would be trapped beneath the ice cap and perish. Chilingarov resurfaced, eight hours later, but it was the mission that he had firmly on his mind as fresh air returned to his lungs. He said: 'We must prove the North Pole is an extension of the Russian landmass ... If a hundred or a thousand years from now someone goes down to where we were, they will see the Russian flag … Our task is to remind the world that Russia is a great Arctic and scientific power.' President Putin personally called each of the members of the expedition to thank them. On 10 January 2008, Chilingarov was awarded the title of Hero of the Russian Federation, one of only 100 people to have been honoured in this way twice in a lifetime.

This was a big moment. It was a clarion call by the Kremlin, which was re-establishing itself as a world power and on a quest to secure oil, gas (20 per cent of which may be in the Arctic), minerals and other coveted and limited resources. It struck a note that was at once an expression of the past, a defiant and determined celebration of the present, and a harbinger of the future.

Fifteen years later, the world found itself at war again, following Russia's invasion of Ukraine. Limited resources and, indeed, resource wars are nothing new. If we ask 'why?', we could find the answer in resources when we chart the origins of so many wars in history. History doesn't end and it may be that it doesn't repeat, but like science, it reveals causes and effects, patterns. These recur. Conflict over resources, most often dressed in the clothes of political or religious ideology, have always and will always recur. Whether or not history repeats itself or recurs, society nonetheless surprises itself and it will probably also never cease to do so.

The 'so what?' of the invasion and the war appeared so clear – the tragic, cruel and unnecessary death, suffering and displacement of millions, the threat to all of the freedoms, memories of the Second World War and fears of a third, an undermining of the security of Western Europe and the Western world, the spectre of chemical or nuclear warfare, the challenge to liberal democracy, stability and a new era of unpredictability, an end of reason. The 'why?' is baffling in what we thought was a post-war era, following which history had ended in the victory of liberal democracy: apparently, we were so wrong as Russia abhors NATO on its doorstep, appears to wish to reinstate the Soviet bloc and holds up 'de-Nazification' as a casus belli to a Jewish leader of a state, de-nuclearized in the 1990s and younger than me.

Commentators were already referring to 2022 as a 'turning point' in history. Russia's foreign minister, Sergei Lavrov, pointed to a 'fateful, epoch-making moment in modern history' in the weeks after Russia's invasion of Ukraine. The President of the European Council, Charles Michel, and Ukraine's president, Volodymyr Zelensky, both hailed as 'historic' the announcement that Ukraine and Moldova were granted EU candidate status in June 2022. 'This is truly a historic moment for Finland, for Sweden and for NATO,' said Secretary-General Jens Stoltenberg after they applied for NATO membership.[2]

So, let's ask the question again. 'Why?' Since at least April 2008, Russia had been told by the USA that the door was open for Ukraine to join NATO. When Russia presented a draft treaty in December 2021 seeking to close the door and push NATO forces back to their post-Cold War positions, backed by the threat of military action, it did not have that effect. But by February 2022, Russia's largest and most valuable export market, Europe, was under threat. Like most other places in the world, oil, gas and coal gas consumption was rising. Exporting gas through the €9.5 billion pipeline to Europe, Nord Stream 2, through Ukraine, was being blocked by the West, particularly its most significant adversary and competitor, the newly energy independent USA. The USA had ceased to need to protect oil and gas supplies from the Middle East, so had withdrawn in an apparently isolationist and nationalist post-COVID retreat. Russia had eyes on China as a replacement market without the ideological strings. Russia had planted a titanium flag at the bottom of the Arctic, opened a pipeline to China and ironically and facetiously exported its first shipment to Boston, Massachusetts, during the freezing winter of 2018. China, like Europe, was an energy importer. Unlike Europe, it ran on 60 per cent coal and 6 per cent natural gas, a relationship that it needed to reverse to protect both its economy and the health of its people and environment. Russia and China signed an agreement including 'limitless' cooperation, weeks before the invasion of Ukraine, the capital city of which, Kiev, is the home of the origin of both Russia and Ukraine, Kievan Rus. It was time for a test.

Circumstances had changed and so had the 'so what?'.

Ukraine hosts substantial gas resources and is endowed with abundant coal. Ukraine has been crucial to the present and future of European energy supply, essential to the transit of Russian gas to the EU, which, until the Russian invasion of Ukraine in 2022, depended on Russia for 40 per cent of its gas supply. Russia had been exporting over 70 per cent of its natural gas to Europe. Most natural gas exported from Russia to Europe was transmitted through territory and pipelines in Ukraine, traditionally a major source of Ukrainian government revenue. Russia invaded Ukraine two years into a five-year deal to use Ukraine's large natural gas transportation system, which is even more valuable so long as the Nord Stream 2 pipeline through the Baltic Sea is mothballed.

Gazprom, previously the Soviet ministry of natural gas and now the largest gas company in the world, controlled the export pipelines and the Russian revenues that came from the exports through Ukraine. Referred to by some analysts as a 'ghost of the Soviet–American Cold War', Gazprom played an instrumental role in the implementation of Russian government policy to gain leverage over Western Europe and latterly in shutting off gas exports to Europe that bypass Ukraine. These actions, such as the restriction of flows to Germany via Poland from the Yamal pipeline, have been referred to by Europe as the 'weaponization' of energy. By July 2022, gas flowing through the TurkStream pipeline to Bulgaria was down 50 per cent and exports through the Yamal pipeline to Poland had stopped entirely, exacerbating what had become a pan-European crisis. Gazprom was curbing exports through all major pipelines to Europe, reducing Europe's ability to build up gas storage ahead of the winter season.

However, there is yet another dimension to consider. Ukraine itself has the second-largest gas reserves in Europe (1.09 trillion cubic metres) after Norway (1.53 trillion cubic metres), most of which are largely untapped and may be accompanied by more that is undiscovered. Although Ukraine lost much of its expertise and capacity to Russia and Eastern Europe in the Soviet era, it has the potential to become energy independent and even a major potential exporter of gas from unused reserves. This would be revolutionary for Ukraine and even for Europe, which is expected to import around 90 per cent of the gas that it consumes by 2030. The industrial heartland of Donbas, an epicentre of the war in Ukraine, and within the five areas annexed by Russia by the autumn of 2022, has one of the largest coal deposits in the world, at around 20 per cent of the total reserves of the former

Soviet Union. (The Donetsk, Luhansk, Kherson and Zaporizhzhia 'oblasts' or regions were annexed by Russia on 30 September 2022. Crimea had been annexed in 2014.) Ukraine's natural resources also include iron ore, manganese, salt, sulphur, graphite, titanium, magnesium, kaolin, nickel, mercury, timber, and arable land. In the next chapter, we explore why these resources matter.

Economic and regulatory reform in Ukraine pre-war made it more attractive for international investment, important to finance the €20 billion needed to develop Ukrainian oil and gas capabilities. Meanwhile, the potential for Ukrainian biogas and its contribution to hydrogen and biomethane markets, in combination with the largest potential gas storage capabilities in Europe (storage and lack of it being vital to European energy security and commodity markets alike), presents Ukraine with the keys to much of Europe's energy future.

Strengthened economic ties with Kiev (or Kyiv),[3] diversification of gas supplies beyond Russia, improved energy security and contribution to hard-to-achieve decarbonization objectives would all be in the best interests of Europe and Ukraine. Not so for Russia. Russia had made very clear its abhorrence of NATO on its borders, as well as the intolerability of Ukraine becoming a member of the EU.

Echoes

There is the echo of history here, recurring if not repeating. Ukraine pledged allegiance to Russia in 1654 and was part of the Soviet Union from its inception in 1922 to when it collapsed in 1991. A revolution was sparked in 2014 when Ukraine was on the brink of signing an association agreement with the EU, mostly related to trade but committing both parties to promote a gradual convergence towards the EU's Common Security and Defence Policy. Viktor Yanukovych, then President, was overthrown and forced to flee to Russia when he refused to sign at the last minute, and Russia instead presented an association with the existing Customs Union of Russia, Belarus and Kazakhstan as an alternative. The new Ukrainian President, Petro Poroshenko, eventually signed the economic part of the European

Union–Ukraine Association Agreement on 27 June 2014, describing this as Ukraine's 'first and most decisive step' towards EU membership. Russia had, in the meantime, annexed Crimea, its only warm-water port, on the Black Sea. 2014 was the 60th anniversary of the year in which Crimea was gifted to Ukraine by Russia under Nikita Khrushchev (1954). This, in turn, was the 300th anniversary of the Treaty of Pereyaslav in 1654, where representatives of the Ukrainian Cossack Hetmanate pledged allegiance to the Tsar of Russia. Russia's invasion of Ukraine in 2022 made it clear that neither history nor territorial claims were fading way.

Energy and Resources behind the War

Europe was already in the grip of an energy crisis, made worse by low wind in 2021 and low storage, leading to record prices for natural gas and electricity, potential blackouts, energy poverty for millions and, contrary to European climate and green policy objectives, increased coal consumption.

In July 2021, the US government had given the go-ahead for the continued construction of Russia's Nord Stream 2 pipeline, waiving sanctions to do so. The Nord Stream 2 project launched in 2011, financed by Russia's Gazprom as to 50 per cent and by a consortium of European energy companies including Shell, Uniper and Engie, most of which have now written off their multi-billion-Euro investments.[4] It would complete in September 2021, with annual capacity of 110 billion cubic metres. It would run under the Baltic Sea avoiding Ukraine, unlike existing Russian supplies, responsible for 40 per cent of Europe's natural gas, which run through Ukraine. Ukraine warned of energy security implications and feared that it would be deprived of transport fees equivalent to approximately 4 per cent of its GDP.

However, the global political landscape turned upside down in a matter of days. Nord Stream 2 had been supported by Germany, which is Europe's largest industrial economy and largely barren of domestic natural gas, albeit in the face of substantial controversy from parties fearing Nord Stream 2 as a political weapon. On 15 February 2022, the Russian State Duma passed a resolution to recognize the Donetsk People's Republic and the Luhansk People's Republic, which was announced by President Putin on 21 February. (The industrial area of the Donbas includes much of Donetsk and Luhansk oblasts.) Germany immediately reversed its support for Nord Stream 2 and suspended certification on 22 February. On 24 February, Russia invaded Ukraine.

Energy and resource competition helps to explain the war in the Ukraine and is a framework for understanding much of its implications. However, this requires joining more disparate dots.

The Russia–Ukraine front was one of several geographically distant but geopolitically interconnected arenas within which resources were being fought over, overtly or quietly, in 2022. These included the South China Sea, the Middle East and Africa, the Arctic and the USA.

THE EAST

On 4 February 2022, China's President Xi Jinping and Russia's President Vladimir Putin met in Beijing, China, to announce a strategic partnership to counter the influence of the USA, with 'no forbidden areas' of cooperation and 'no limits'. (Echoes of 'no limits' reverberated a year later in March 2023, when Messrs Putin and Xi met in Moscow to sign an agreement on 'deepening the comprehensive partnership' between their two countries.)

Russia supported China's claims to resource-rich Taiwan. China supported Russia's resistance to NATO enlargement and demands for security guarantees over Ukraine, and meanwhile conducted military exercises expressing its claims over Taiwan and the resources of the South China Sea. Russia and China pledged to work together on space, climate change, artificial intelligence and the internet. President Putin used the occasion to celebrate a new gas deal with China worth US$117.5 billion and promised to ramp up exports to the Far East. Since President Putin came to power in 1999, Russia had become China's top energy supplier. The meeting took place on the eve of the Winter Olympics. Russia invaded Ukraine as the Games ended.

China and Russia signed the 'Treaty of Good-Neighborliness and Friendly Cooperation' in 2001, the first pact that they had signed since 1950. It said that China and Russia would 'remain friends forever and never become enemies'. By 2012, China was Russia's largest trading partner. Western sanctions after Russia's invasion of Crimea in 2014 had strengthened the relationship between China and Russia, including a series of energy agreements such as the 'Power of Siberia' deal, signed three months after the annexation, which helped Russia rebound. By 2018, President Xi Jinping called Russian President Vladimir Putin his 'best friend'. In September 2022, President Xi made his first trip overseas since the start of the COVID-19 pandemic. He travelled to Uzbekistan to meet Vladimir Putin at the Shanghai Cooperation Organisation (SCO) Summit. Established in 2001 and including India, Pakistan and Iran as members alongside four ex-Soviet Asian states and with Turkey as an observer, China and Russia had long sought to position the SCO as an alternative to Western multilateral groups.

Recently, in January of 2022, the USA had withdrawn its support for the EastMed pipeline, designed to bring natural gas from Israel to Europe. The €6 billion EastMed pipeline project, jointly developed by the Greek utility DEPA and the Italian utility Edison, is to reach 1,900 kilometres undersea from Israel to Southern Europe, with capacity of 10 billion cubic metres per year, and is set to complete in 2025. It could ease Europe's dependence on Russia and Turkey. The USA's withdrawal of support was at least partly attributed to an attempt to avoid offending Turkey, which has claims over natural gas in the Mediterranean. President Erdogan made no bones about it, stating that 'it will only happen through Turkey'.[5] Months later, once Europe's dependence on Russian gas was fully exposed, this carried many hallmarks of a major policy backfire.

THE WEST

But it was partly because, just as the flows of energy were changing in the East, the dynamics of the energy system had also been changing in the West. The USA had explained that the reason for its withdrawal from EastMed in January 2022 was that 'We are shifting our focus to electricity interconnectors that can support both gas and renewable energy sources.'[6] In parallel, the USA's new-found energy independence since 2019, thanks to the revolutionary implementation of fracking technology, has made it a net exporter of oil and gas. In 2008, the USA spent almost $400 billion on oil imports. In 2020 it spent nothing. The USA had become the world's largest consumer of oil, as well as the world's largest producer of oil at over 11 million barrels a day, followed by Russia and then Saudi Arabia, which produces more than double that of the next largest oil-producing nation. The USA has also become the world's largest producer of natural gas at around 950 billion cubic metres per annum, then Russia at just under 700 billion cubic metres, then Iran at around 250 billion cubic metres.

Demand for and dependence on oil had defined the relationship of the USA and its allies with the Middle East following the Second World War. Much of the fuel for the social, religious and political conflicts that define the tensions in the region finds origins in the spheres of influence agreed between the UK and France, with the assent of the Russian Empire and regarding control and partition of what had been the Ottoman Empire in the First World War. (The Sykes–Picot Agreement in 2016, a political settlement designed in Europe, together with the associated 'Sykes–Picot' line, a new border or 'edge' between British- and French- controlled territories reaching from the northern mountains of Iraq to the Mediterranean coast of what is now Israel, in many ways established the basis for much of the regional resentment of the West and the unrest of the twentieth century.)

A combination of the discovery of much of the world's oil and gas resources, together with increasing global demand and dependence, placed the Middle East at the heart of the answer to the world's energy needs. In particular, the major owners of resources, Saudi Arabia, Iran and Iraq, have found themselves at the intersection of growing rifts between Sunnis and Shiites, political instability and both domestic and international conflict. The prospect of energy independence, or at least reduced dependence on the Middle East, had a profound influence on the reduced interest from the USA in continued investment in military operations from 2020. Other priorities and frontlines were appearing. Energy policy needed to change to keep up, but with change came the potential for backfire.

An illustration of this potential for energy policy backfire had slowly been playing out in the USA. The cancellation of the Presidential permit for the Keystone XL pipeline on environmental and climate grounds in June 2021 had halted plans to bring oil from Canada and North Dakota to the Gulf States. At the same time, Russia was a major supplier of oil to the USA, capturing the number two spot in 2021 and competing with Mexico. While Russia was responsible for only 3 per cent of crude oil imports, its share of petroleum product imports was 20 per cent. Russia was the third largest oil producer in the world after the USA and Saudi Arabia. The International Energy Agency (IEA) data suggests that it was the world's largest exporter of oil and petroleum products (although the US Energy Information Administration claimed that the USA was larger).[7] Russia's output includes naphtha (used as a solvent), vacuum gas oil (used to increase petrol output from refineries), gas oil, also known as red diesel (used in farm machinery), and fuel oil (for home and industrial heating boilers).

On 8 March 2022, the USA banned the import of Russian oil, liquefied natural gas and coal after Russia's invasion of Ukraine. The impact on prices and the cost to the USA of its energy policy and reliance of Russian imports would be felt at every level, from the petrol pump to production facilities to high politics. The effect of the Russian invasion of Ukraine and the extent of European reliance on Russian oil and gas, which was substantially higher still and which had fuelled European industrial growth for 25 years, would be nothing short of a series of simultaneous all-out energy, economic, social, environmental and political crises in Europe. But the super-tanker of Russian energy influence had been moving slowly and deliberately for many years and much about the chronicle had been foretold.

THE NORTH

On 28 January 2018, a Russian tanker carrying liquefied natural gas from Russia's Arctic had arrived in Boston, Massachusetts, in a particularly cold winter and given a shortage of pipeline capacity from gas-rich Pennsylvania. The tanker arrived at a terminal owned by Nord Stream 2 consortium partner, Engie. It had come from the Yamal facility, a US$27 billion project, majority owned by Novatek, Russia's largest independent producer of natural gas. The US Treasury Department had issued sanctions aimed at weakening Russia's energy sector in July 2014 after Russia's last invasion of Ukraine, involving the annexation of Crimea and the Russian backing of separatists in eastern Ukraine. The sanctions forbade any financing for projects belonging to Novatek. The loophole was that the sanctions did not prohibit the purchase of natural gas that originated from Yamal, and the trail was at best blurred. The *Chris De Margerie*, an ice-breaker named after the late former CEO of French energy giant Total (which owned 20 per cent of the Yamal project), had picked up the shipment on 9 December 2017. The loading of the tanker had been overseen personally by Vladimir Putin. The cargo was dropped off at the National Grid-owned Isle of Grain liquefied natural gas (LNG) facility near London on 28 December. The Engie tanker picked it up on 30 December. On a particularly cold Sunday evening in January, after having first been diverted to Spain and at one point sold to the Malaysian energy giant Petronas, the blue-hulled tanker arrived at an LNG terminal on the Mystic River in Boston.

This event in the USA was the outcome of geopolitical shifts that had taken place far away, both in physical distance and time, in another arena, the Arctic. And yet another dimension, climate change, had meanwhile been turning up the heat on the relationship between Russia, Europe and the USA, and threatened to be a catalyst for conflict. The Arctic had been warming three to four times faster than the rest of the world. In 2020, minimum sea-ice cover was around 2.6 million square kilometres smaller than the 1981–2010 average. Melting ice opened a shipping corridor from the Bering Strait (between Siberia and Alaska) and the Barents Sea, allowing access for over 1,000 cargo ships in 2020. The shipping route is controlled by Russia, which charges a fee for passage, but the same ice that is melting to open the shipping route is also used to protect Russia's northeast coast, making it more vulnerable. (In the run up to a NATO summit in Brussels in June 2021, NATO said that melting ice 'could lead to new geopolitical tensions'.[8])

Russia, Canada and Denmark all lay claim to the Lomonosov Ridge. (In December 2013, Canada issued a passport to 'Santa Claus' and 'Mrs Claus'. On Christmas Day, *The Washington Post* declared: 'Canada just enlisted Santa Claus in its effort to control the Arctic ... It's no joke.') This claim involves rights over supplies of 10 billion tonnes of hydrocarbons. As the ice cap is melting, exploration and drilling for oil and gas are getting easier. As the 2007 expedition's leader, Artur Chilingarov, had said, his mission had been to prove that 'the Arctic is Russian'. [9]

In May 2022, three months after Russia's invasion of Ukraine, Sergei Lavrov, Russia's foreign minister, unequivocally re-stated the claim: 'It has been absolutely clear for everyone for a long time that this is our territory.'[10] Fifty-three per cent of the Arctic coastline is Russian land. Since 2007, *The Economist* observes that at least 50 'Soviet era' military outposts have reopened, and that Russia has built at least 475 military sites along its northern border since 2016.[11] In August 2022, the USA announced that it was planning to appoint an Artic Ambassador.

The Arctic Ocean is about 3 per cent of the world's surface area, about the same size as mainland Russia. It is estimated to contain 22 per cent of the Earth's oil and natural gas, together with large quantities of minerals, a treasure that might test any country's climate change credentials. Russia set to work to extract oil and gas from Yamal.

As if any complication were needed, China claims to be a 'near Arctic' nation. As a signatory to the Treaty of Svaldbard in 1925, it claims the same rights as Norway to exploit the archipelago of the same name. In 2010, Chinese Rear Admiral Yin Zhuo claimed that the 'Arctic belongs to all the people around the world'.[12] This has the potential to frustrate Russia, but Russia depends for its bid for the Arctic, as it does in many other aspects of its economy, on Chinese finance. China has a 30-year contract with Russia to import gas from the Yamal fields. In 2016, the Silk Road Fund and China National Petroleum took a 30 per cent stake in Yamal LNG, the liquefied natural gas plant in Sabetta, Yamal, alongside France's Total, as part of its plans for a 'polar silk road' in which it has invested more than US$90 million. Commercial operations were launched on 8 December 2017 by President Putin in the presence of Saudi Arabia's energy minister, Khalid al-Falih.

It was in this context, symbolically and ironically, that one of the first shipments landed in the USA, that cold Sunday night in Boston, Massachusetts, on 28 January 2018.

Shipments had also commenced to China, in conjunction with gas pipelines running from Siberia to China. As part of Russia's 'Pivot to the East', on 18 December 2021, President Putin discussed Gazprom's Power of Siberia 2 mega-pipeline across Mongolia, with a capacity of 50 billion cubic metres, with President Xi of China during a video conference. It had been given the go-ahead in March 2021 by President Putin to complement Power of Siberia 1, which transports gas from Russia's Chayandinskoye field to northern China. Chinese demand for gas is expected to double by 2035, according to consultants such as McKinsey, and so, argued analysts at the European Council on Foreign Relations, would give Russia additional leverage in pricing negotiations with Europe, boost China as an alternative market for gas and, indeed, reduce financial risks for Russia associated with supplying Europe.

The 4,107-kilometre Yamal–Europe pipeline provides much of Europe's natural gas into Poland and Germany via Belarus. When Russia announced suspension of supplies to Poland and Bulgaria in April 2022, European Commission President Ursula von der Leyen said that the suspension was tantamount to 'blackmail'. [13] When Germany faced a 60 per cent reduction in gas supplies from Russia in June 2022, it hailed a 'gas crisis'. Robert Habeck, economy minister, said gas was being deployed 'as a weapon against Germany'.[14] Germany had been diversifying to reduce dependency on Russian gas imports from 55 per cent in February 2022 to 35 per cent by May 2022, hurriedly ordering new LNG facilities and staring at gas storage facilities left empty by their owner, Gazprom, since the beginning of the year. Prices rose and hearts sank as the plot thickened. By July, the IEA was warning that after Canada had repaired and returned Siemens turbines needed to get Nord Stream 1 up to full capacity again, Russia might turn off the taps for good. When Russia squeezed supplies in July 2022, as

Gazprom announced that it was reducing gas flows into Germany by 20 per cent to allow work on the turbine, Ukraine accused Russia of waging a 'gas war' against Europe. By August, supplies had been cut completely.

The Energy Security Problem

While events such as COP26 in Glasgow in December 2021 offered the promise of an alignment of countries towards decarbonization and sustainability, the realities of the economic, political and energy system present hard challenges to hopeful rhetoric. The choices available to Europe in replacing the 150–190 billion cubic metres, or 40–45 per cent of the natural gas for which it depends on Russia, included Norway, Britain, the Netherlands and Azerbaijan, via pipelines and shipments of LNG from Qatar, and the USA. The European Commission estimated that the USA and Qatar could replace 60 billion cubic metres of the gas that it gets from Russia and that new wind and solar projects could replace only 20 billion cubic metres of gas demand (about 10–15 per cent of the need). Solutions to replace Russian oil were so scarce that it took until the end of May 2022 to enforce a partial ban. Biomethane and hydrogen could help by 2030, as could an initiative to triple renewable energy capacity by 2030, but these are solutions at scale for the end of the decade. Energy storage was nowhere near the scale needed to balance intermittent renewables – batteries can store only a minute amount of Europe's energy needs. Meanwhile, nuclear availability was falling in Britain, France, Germany, and Belgium as ageing plants faced outages or were decommissioned. Europe would instead rely on fossil fuels from the USA.

Beside the political instability, the horror of war and violence, and the displacement and injury of millions of innocent people, the war in Ukraine signified a profound dislocation, a hand-brake turn for energy and foreign policy that shone a global-scale spotlight on the issues of energy security and the economic and environmental consequences of failure for countries, companies, and populations.

Russia and Ukraine held the keys to other critical resources such as nickel and zinc, essential for many of the technologies needed to diversify energy production and, of course, food.

Spikes in wholesale energy prices (gas prices in the Netherlands were eight times the normal level in the summer of 2022 and power for 2023 was being traded at six times the five-year average in Germany) and limited supplies drove inflation, curtailed GDP, halted manufacturing production and increased energy poverty, with delivered energy prices for businesses and individuals more than doubling in some countries, more than quadrupling in others, crushing margins for industry and leading ordinary families to choose between food, heat and fuel. Fear gripped governments and markets. Memories of the 1970s flashed back. Between 1973 and 1975, the global economic growth rate had dropped around 90 per cent after the price of oil increased over four times. The International Energy Agency declared 2022 the world's first 'truly global energy crisis'.[15]

Energy security, a familiar problem for the USA, beset perennially by severe weather events that destabilize the grid, had very suddenly hit Europe hard, and it was unprepared. Twenty-five years of post-Cold War energy market liberalization in Europe had fuelled growth that depended on unlimited access to cheap Russian gas, particularly in Germany. Germany had decided to close all of its nuclear power plants in favour of intermittent renewable power, while France's nuclear fleet, which had exported to its European neighbours, had aged into a fleet of faulty reactors. Now the Russian gas taps were turned the wrong way. While oil and gas companies reported record profits, utilities struggled to survive. In Germany, talks started about bailing out Uniper SE (as Uniper was receiving only 40 per cent of its Russian gas orders, it was costing the company US$30 million a day), France started looking at nationalizing Electricité de France SA (in the first weeks of July 2022, both Uniper and EDF were under preparation for nationalization) and the UK had to take control of Bulb Energy Ltd. France could no longer depend on an ageing fleet of nuclear reactors (in April 2022, EDF confirmed that half of the French nuclear fleet of 56 reactors had been taken offline) and was turning to the import markets. But so was almost everyone else in Europe. Governments stepped in to replenish storage sites at whatever cost. Germany had to seize and rescue a former unit of Gazprom as it owned 20 per cent of the country's storage capacity, at the same time as bailing out Wingas, a key supplier to business, and passing a law to give the government the ability to seize critical energy infrastructure. But they also had to step in to relieve the energy-cost crisis and the cost-of-living crisis that it caused. The German government started distributing payments to households. France planned to 'double down' on the €25 billion of spending and tax cuts to shield consumers and business. Italy set aside nearly €40 billion to subsidize energy bills, the UK £37 billion.

By June, Germany feared that the Russian gas taps would be turned all the way, to off. While Russian gas represented a crucial one-third of European supply, it was only 2 per cent of Russian GDP. The pain in reducing supplies from Russia to Europe would be felt far worse by Europe than Russia. Efforts to build storage in defence had been extraordinarily effective, increasing from 26 per cent in March to over 50 per cent by June, but had hit limits as reserves were drawn on to power air conditioning in the sweltering summer, raising fresh concerns about a cold winter. Concerns that Russia was deliberately squeezing supply were not isolated to the short term. Prices for delivery of natural gas the following winter reached four times the normal level. What *The Economist* referred to as a 'gastastrophe' was threatening to upend European stability and unity and was being compared to the euro crisis in the early 2010s and the migrant crisis in 2015.[16] By the summer of 2022, European economic ministers were meeting to prepare 'solidarity' agreements to provide emergency supplies of fuel to distressed member states as a 'last resort' (rather than ban exports) if Russia entirely halted gas supplies to the European Union.

At the time, only six 'solidarity' agreements existed across the EU. Not all governments were that keen. Poland had objected to Nord Stream 1 and Portugal had felt frustrated by what it saw as a refusal by Germany to explore alternatives to Russian gas in the past, leaving the Iberian Peninsula with a dearth of pipelines with the rest of Europe. Eastern Europeans resented Germany's lobbying for an exemption to Russian sanctions to return a gas turbine serviced in Canada to Russia to re-start the Nord Stream 1 pipeline following maintenance. German gas companies had objected to the European Commission's proposals to combine European purchasing power to make joint gas purchases, fearing competition. If Russia turned off the gas taps, emergency supplies would have to be pumped from North Africa and Italian LNG ports, from Spain and France, and from Norway. As fears of winter shortages of 20–30 per cent in central and Eastern Europe grew, so did fears for European unity.

A halt to Russian gas flows might cost 2–4 per cent of GDP and add 2.7 per cent to inflation. Germany had no LNG regasification facilities in 2022. The government acquired floating terminals and started fast-tracking approvals, with operations starting in 2022. Meanwhile in the three months after Russia's invasion of Ukraine, around 15 per cent of the LNG entering Europe to replace piped Russian gas was estimated to have come from Russia. Energy Intelligece Group estimated that Russia earned nearly US$400 million a day from sales of piped and frozen gas to Europe, considerably more than Europe was spending to defend Ukraine.

While Europe hurriedly sought to diversify sources of gas, turning to LNG, Europe might pull together, but it might also pull apart. Shortages could create competition between countries for the same limited resource and even reduce support for standing up to Putin.

The Climate and Environment Problem

The energy crisis faced by Europe caused a setback not just for its energy security, its energy markets, its suppliers, its system and the cost of living, but also for climate and environmental policy. Lack of storage and low production levels from intermittent sources such as wind had already conspired to drive up gas prices in 2021. Coal-fired power stations in Germany, Ireland and the UK had been fired back up. In sharp contrast with the conclusion of the COP26 climate talks only eight months earlier in November 2021, in which the then UK Prime Minister, Boris Johnson, had declared that the world had reached the point of no return in phasing out coal, Ukrainian Foreign Minister Dmytro Kuleba was, at the same time, warning Europe that Russia was building up forces near its border. Soon after the invasion, Germany stated that it could extend the life of coal or nuclear plants to cut the reliance on Russian gas.

By June, after Russia cut capacity on Nord Stream 1, the main gas export pipeline to Germany, by 60 per cent, emergency laws were being passed in Germany to reopen up to 10 GW of idle coal-fired power plants for electricity generation, increasing capacity by a third (at odds with its climate policy to phase out coal by 2030) and gas supplies (55 per cent from Russia) were being auctioned to industry to curb consumption. At the end of August 2022, Nord Stream 1 supplies were completely shut down by Russia, for 'repairs'. On 26 September 2022, the day before the 'Baltic Pipe' was being opened to send natural gas to Poland via Denmark from Norway, pressures in Nord Stream 1 and Nord Stream 2 dropped to almost zero following underwater explosions and gas leaks.

The German finance minister also rejected EU plans for a de facto ban on the sale of new combustion engine cars by 2025, a pillar pledge from COP26. The same week, the Netherlands lifted capacity restrictions on coal that it had imposed earlier in the year until 2024, following in the footsteps of Germany and Austria. The UK had already decided to keep a reserve of coal-fired power plants available, as opposed to original plans to close most of them within months. Germany and Britain looked to increase domestic gas production as well as diversify imports. By the end of June, the UK had appointed a 'czar' to ensure that it could keep the lights on for the winter as the energy crisis deepened. The Irish Green Party supported new gas generation on the grid to balance and enable intermittent renewables, support inertia, and provide baseload power. The G7 endorsed gas, tolerated coal (in terms of resource scarcity, at 2020 levels of production, according to scientist, policy analyst and author of *How the World Really Works* Vaclav Smil, coal reserves would last for around another 120 years and gas reserves for around another 50 years) and encouraged low-carbon not just renewably sourced hydrogen.

At the end of June, the US Supreme Court clipped the wings of the Environmental Protection Agency, limiting its ability to curtail emissions in energy states. In July 2022, when the European Commission's proposal to classify nuclear power and natural gas as 'green' energy was approved by the European Parliament, *The New York Times* observed that this was 'likely to reverberate far beyond Europe's borders and set a benchmark that could be replicated around the world'.[17] As the crisis deepened, the numbers became clearer still. The IEA forecast that European coal consumption would increase 7 per cent in the year, driving global coal consumption to match the record levels of 2013. In the short term, as natural gas and coal demand continues to increase and alternative energies are, as yet, unable to meet the shortfall, fossil fuel consumption, unabated, is set to rise rather than fall.

To make things worse, as the air conditioners in Europe cranked up during the record-breaking and blistering heatwaves of the summer of 2022, the rise in energy demand pushed the European energy system to breaking point, with lower-carbon generators such as nuclear and hydro amongst the first to buckle as they struggled with water resources. The environment suffered with the generators. The French regulator waived caps on heated water discharge into rivers, damaging the environment. High temperatures and low rainfall in Germany drained the Rhine River, restricting coal supply to power plants. Meanwhile, the high pressure on the balmy hot summer days produced less wind. Heat reduced the efficiency of gas and solar power plants alike, while dried- up reservoirs reduced hydro production by nearly half in France at times. Before the end of July, the citizens of Hanover in Germany were having cold showers, as energy was curtailed. Public fountains and night-time lights on major public buildings were switched off. By August, German manufacturers started halting production.

By September, there was a denouement. Europe was looking with trepidation beyond its stores of natural gas, perhaps enough for that winter, but not for the next. It was doing all it could to add sources of imported natural gas to diversify from Russia, having reduced Russia's share of imported gas from 40 per cent to 10 per cent. The President of the European Commission, Ursula von der Leyen, said that European member states had been able to stockpile gas reserves for the winter to 84 per cent of capacity by September, ahead of the deadline that had been set for October. The USA, Norway and Algeria were named as 'reliable' gas suppliers. The UK was the first to pay £150 billion to cap the price of energy for consumers and businesses but other European states started to follow. In the face of limited supply and preferring to cap and tax the energy supplies rather than subsidize the users, the European Commission was turning to the demand side. In the summer it had called on member states to reduce gas use by 15 per cent. In September, it called for peak electricity use to be cut by at least 5 per cent.

By November, there were calls for a 'Green Marshall Plan' in Europe to switch from Russian oil and gas imports and to create a green-energy economy for Ukraine.

A Web of Interdependence

The Russian invasion of Ukraine laid bare an intricate web of interdependence that had been quietly tying together the European energy system and fuelling its growth for more than two decades. The war, and the dislocation of critical supply chains and pricing levels that this created, blew apart much of the basis for the balance of power and economy since the Cold War. During this time, the USA had finally achieved one of its most strategic objectives following the Second World War, energy independence, or at least self-sufficiency. However, it remained intimately connected to Europe and the rest of the world through global markets, where energy prices had soared, through trade, and through geopolitics. In the most recent years under President Trump, the USA had progressively sought to withdraw from the Middle East. Now under President Biden, it claimed that it was back, to try to restore balance and manage the interface between resources and the economy and navigate the sharp edges of geopolitics. Would energy independence deliver energy security?

Energy security had even been a key attraction of some popular decarbonization initiatives, such as the electrification of transport, which promised to reduce dependence on oil. Indeed, when Senator Joe Manchin finally backed President Joe Biden's US$369 billion climate bill (reduced from US$555 billion) in late July 2022, he explained it as an action to protect both US oil and gas interests as well as climate and clean energy interests. 'You have to have energy security, you have to be energy independent if you want to be a superpower in the world ... That's what China does and that's what Russia has had.'[18] The legislation was renamed from 'Build Back Better' to the 'Inflation Reduction Act'. It was the largest single proposed investment in clean energy and climate programmes in the history of the USA. Opposed by the Republican Party, it succeeded only with Democratic Party backing to help the USA become energy self-reliant in the future. At the same time, the sheer scale at which the USA would stand to benefit by feeding Europe's need for natural gas in the post Russia–Ukraine world was becoming clear. Europe had no choice but to diversify its sourcing of natural gas and the USA would replace Russian supplies. Prices for natural gas were, at the time, around five times higher in Europe than in the USA.

The realities of the present accompanied hopes for the future and were a constant reminder of the prevailing inescapable interconnections of the time. In July 2022, President Biden had been back to Saudi Arabia, looking past the diplomatic crisis following the assassination of Jamal Khashoggi, a *Washington Post* journalist, ostensibly holding regional nuclear talks, developing the rapprochement between Israel and its Arab neighbours, and seeking to contain Iran. The big ask of Saudi Arabia was to increase oil supplies to help reduce the soaring costs of oil to the West, while Russia diverted its exports, slammed by Western sanctions, to China and India in the East at discounted prices. President Biden's visit to Saudi Arabia created uproar and consternation. In his defence, US Democratic Congressman Brad Sherman said: 'The price of oil means people die in poor countries. It raises the price of food and fertilizer and it means people die by the hundreds of thousands, not just from starvation but also from the disease the malnourished tend to acquire.'[19] At that point in time, only Saudi Arabia exported more crude oil globally than Russia, in fact by a factor of two. Canada, Iraq and the United Arab Emirates comprised the remainder of the top five oil exporters.[20] Iran's exports, given sanctions, were flowing most to China – at a discount.

Understanding the Limits

The challenge of decarbonization to prevent the worst effects of global climate change was daunting before the war in Ukraine. Carbon emissions needed to peak by the middle of the 2020s if the global average surface temperature of the Earth were to be kept below 1.5°C, the level beyond which runaway climate change risks creating irreversible damage, according to the consensus of United Nations scientists. The increasing demand for fossil fuels and the imperative to find new sources of conventional supply will make this challenge even greater.

Organizations from the International Energy Agency to financial analysts at Goldman Sachs have published pathways to achieve outcomes aligned with a 1.5°C or 2.0°C implied temperature rise and net zero carbon emissions by 2050. The Glasgow Climate Agreement signed at COP26 in December 2021 involved commitments by most of the world's governments. The 'net zero' targets that have now been set by almost all major governments and companies involve removing as much excess carbon as is generated. Sequestration through natural sources such as forests is part of the answer, along with substantially smarter and more carbon efficient solutions to agriculture, construction, manufacturing, transport, and energy management. Reducing demand through increasing efficiency and productivity has to be a key – indeed it is arguably the largest and cheapest source of greenhouse gas emission reductions, or what the International Energy Agency has referred to as the 'first fuel'. But most global commitments, more than reducing demand through efficiency, involve reducing carbon emissions by creating new sources of renewable energy at large scale. COP27 in November 2022 in Egypt did not substantially change the direction of travel.

While there is no such thing as zero carbon energy production, the largest existing, proven scalable and replicable sources of renewable electricity are wind, solar, hydro, biomass and geothermal. Intermittent generation from wind and solar can be addressed by pairing with batteries. Green gases such as biomethane and hydrogen are part of the solution for industry and transport, along with ammonia. Nuclear can generate power with low or no net marginal carbon footprint. In many countries, low-carbon and renewable technologies with the ability to generate at massive scale have reached a price of production that is competitive with the grid. However, their manufacture requires resources, and time. And there are limits, both for the planet and for people, and the geopolitical ramifications resound.

- Russia's interest in Ukraine is intrinsically linked to access to and transport of energy and other natural resources.
- Competition for resources is creating security crises in multiple arenas, which are interconnected and interdependent.
- Competition for resources defines security crises and how conflicts progress or conclude. Demand for resources also defines the global energy and climate crisis, how it is deepening and how it might be addressed.

CHAPTER 2

Testing the Limits of the World's Resources

• • •

'Earth provides enough to satisfy every man's need, but not every man's greed.'

Mahatma Gandhi

- Competition for resources is creating security flash-points on multiple fronts.
- The resources that are being fought over are being consumed at an unsustainable rate, causing an environmental and climate crisis at the same time as a security crisis.

On 8 March 2022, the London Metal Exchange (LME) nearly collapsed in chaos over nickel. Since Russia had invaded Ukraine a few days before, on 24 February, global commodity markets had been in turmoil. LNG orders were paused, financing for trade in raw materials evaporated and sales of Black Sea wheat froze. Russia exports around 5 per cent of the world's nickel. Between Monday 7 and Tuesday 8 March, the price of nickel more than doubled. The LME suspended trading in nickel for a week and cancelled US$3.9 billion in trades, because, as the CEO put it, prices were 'becoming disconnected' from 'physical reality'.[1]

The LME is key to global trade in critical metals such as copper, aluminium and nickel. It sets the global reference prices for these commodities and trades around half of the total world volume in futures contracts, worth some US$60 billion a day. It was acquired in 2012 by Hong Kong Exchanges and Clearing Ltd. In addition to its electronic market, trades happen at 'the Ring', one of the world's last remaining 'open-outcry' trading floors, where brokers shout orders at each other, a short distance away from the circle of sawdust around which traders first gathered in the Jerusalem Coffee House in the City of London in the early nineteenth century. Today, mining companies use the market to offset price risks and the financial sector uses it to invest or speculate. The LME keeps the actual metals that are being traded through the financial contracts in a network of warehouses, so that they are physically backed.

Behind the crisis for the LME was a 'short' trade by one of the world's largest producers of nickel and stainless steel, Tsingshan Holding Group Co, closely held by a Chinese tycoon, Xiang Guangda. Tsingshan had bet that the price of nickel would fall because of rapid increases in production from new factories in Indonesia. As the potential supply squeeze of nickel from Russia hit home, so the price of nickel rose and caused a 'short squeeze', where contracts for nickel had to be bought back rapidly to settle the bets. This sent prices higher still and, on 8 March 2022, the price of nickel had spiked 250 per cent overnight. The LME cancelled billions of trades, effectively re-winding the market to the closing price on 7 March. This bailed out Tsingshan and its banks but wiped out the other side of the trades, made by investors that had placed the opposite bets, and sparking lawsuits for hundreds of millions of dollars from the hedge funds that had made them. The debacle also drew scrutiny and criticism from the International Monetary Fund. Nickel is not only the key ingredient in stainless steel used by all of us every day but is also key to the energy transition beyond fossil fuels, used for instance in electric car batteries.

The LME's crisis marked a turning point in the metals market and a new era where 'mineral security' would join 'energy security' at the top table of geopolitics.

Resource Scarcity

War, in a brutal form that we can see and recognize and abhor, rages over fossil fuel energy resources in Eastern Europe. Meanwhile war, in other forms, quieter and bloodless but just as deliberate, is being waged between the Far East and the West over other resources, including minerals, many of which are critical to renewable energy and the transition to a lower-carbon global economy. These resources, as well as other necessities such as water and food discussed later, are being stretched to their limits, to the edge.

Most of the materials and metals or 'rare earths' required for a lower-carbon global economy, from renewable energy to electric vehicles, are themselves limited, are sourced from a small number of countries, namely Africa and Russia, and most in turn are processed in China, with which the USA is in a spiralling trade war.

Taiwan, the source of most of the world's computer chips, the subject of a global shortage with massive implications for supply chains from automotive and consumer, to energy and military markets, is in many ways China's Ukraine. China has never accepted its independence and the prospect of military action by China to subsume Taiwan resounds. When speaker of the USA House of Representatives, Nancy Pelosi, visited Taiwan in August 2022, Russian Foreign Minister Sergei Lavrov drew the comparison when he told the media that 'it reflects the line we have been talking about in relation to the Ukrainian situation'.[2]

As Russia complains about the prospect of NATO at its borders, China's claims over the South China Sea, patrolled by the USA and other NATO countries, echo both in its claims of historical context and in geopolitical tensions over energy and other limited resources. The South China Sea, territory claimed by China around a 'nine-dash line' on a map, is the conduit for most of the world's transported oil and liquefied natural gas, produced in the Middle East and Southeast Asia and destined for China, its rival Japan and other key Asian import markets. It is itself a vast source of resources. The South China Sea is a potential military flashpoint, where mistakes are feared as much as any economic competition or conflict over resources.

At the same time, back in Russia, sanctions imposed by the USA, Europe and other countries after the invasion of Ukraine targeted key Russian industries, including financial services. One company and its oligarch spared sanctions was Norilsk Nickel, the US$50 billion Siberian miner. The company plays a crucial role in the global metals market. The US Treasury still has painful memories of market turmoil sparked by placing aluminium producer Rusal under sanctions in 2018. Norilsk produces 15 per cent of the world's high-grade nickel used in batteries and 40 per cent of its palladium. The concern was that sanctions could make the markets for both metals collapse, hitting supplies needed for cars and microchips.[3]

The USA's reluctance to cut back on Russian nickel and palladium mirrored Germany's and other European countries' reluctance to cut back on Russian natural gas and oil until alternatives are found. In the meantime, Russia derived the revenues it needed to finance the war that created the crisis. A vicious circle.

Mineral Security

Renewable energy systems such as wind turbines, power transmission systems, batteries, electric cars, and their supply chains all depend on key minerals that are extracted from the earth by mining, quarrying and pumping. Demand for copper, lithium, nickel, manganese and cobalt is expanding rapidly worldwide. The International Energy Agency has predicted that the 2050 net zero target will 'supercharge demand for critical minerals' in the transition from a 'fuel intensive to a mineral intensive energy system'.[4] Rapidly increasing demand and limited supply can result in increasing prices and shortages.

The environmental and social impact of – and emissions from – mining will be a key concern for some. Problems will include demand for already scarce water, harm to biodiversity, impacts on land use, communities, child labour and the generation of waste. But another concern will be competition and potential conflict over mineral resources.

Resource demand is already on the edge, according to the International Resource Panel (IRP), a global scientific panel hosted by the United Nations Environment Programme. In 1900 the world consumed 7 billion tonnes of primary materials, but this reached 90 billion tonnes in 2017 and is expected to reach 186 billion tonnes by 2050. Accelerating demand for resources has already breached four of nine planetary boundaries, increasing the chances of irreversibly changing major Earth systems, including the climate, and destroying biodiversity. Demand is not balanced with efficiency or productivity. Less than a third of existing metals have a recycling rate above 50 per cent; most specialty metals have a recycling rate of less than 1 per cent.

The supply of mineral resources required for the low-carbon economy is highly concentrated. Concentration of supply of the minerals needed for net zero is much higher than for conventional energy. Currently, over 80 per cent of lithium supply comes from the top three producers. The top three producers of oil represent around only 30 per cent of supply, and 95 per cent of Europe's rare earths, needed both for electric cars and wind turbines, comes from China. Moreover, 60 per cent of Europe's cobalt originates from the Democratic Republic of Congo and 80 per cent of Europe's imports are refined in China. China, which produces almost 70 per cent of the world's solar panels, some 60 per cent of the polysilicon and almost 95 per cent of the solar wafers from which they are made, has already secured an 80 per cent global market share in battery cell manufacturing capability. The US government has called for 'battery security' by making manufacturing domestically a strategic priority. The European Union has launched a European Battery Alliance, identified 30 'critical raw minerals' and warned in 2020 that 'access to resources is a strategic security question for Europe's ambition to deliver the Green Deal'.[5] By mid-2022, the European Commission was drawing up plans to boost mining and production of critical materials such as lithium, cobalt and graphite. It expected a five times increase in demand for rare earths for wind turbines by 2030, a 15 times increase in the demand for cobalt and graphite and a 60 times increase in lithium by 2050. The USA's Inflation Reduction Act of 2022 promoted domestic manufacturing and supply chain for clean energy technology. The European Commission followed suit with the Green Deal Industrial Plan. In

March 2023 it introduced the Critical Raw Materials Act, aimed at breaking Europe's reliance on Russia and China for metals and minerals required for energy transition, defence, aerospace, digital and energy security, and the Net Zero Industry Act, aimed at increasing clean energy manufacturing capacity to at least 40 per cent of its generation needs. Given that this target was double prevailing capacity, it was accompanied by fears of supply chain constraints and that it might drive cost increases. Political efforts to break dependence on Chinese supply chains, on the one hand, and to reignite industrialization in Europe and the United States on the other invited concerns about slowing the pace of the clean energy transition and creating inflationary pressures.

In 2017, the World Bank released a report that modelled the mineral extraction required to generate about seven terawatt hours of electricity, enough for half of the global economy, with solar and wind by 2050.[6] (Wind turbines can use up to nine times more minerals than a conventional gas power plant (typically requiring 1,500 tonnes of iron, 2,500 tonnes of concrete and 4,500 tonnes of plastic.) The World Bank report estimated that this would take 17 million metric tonnes of copper, 20 million tonnes of lead, 25 million tonnes of zinc, 81 million tonnes of aluminium and 2.4 billion tonnes of iron. For solar panels, extraction of neodymium will need to rise

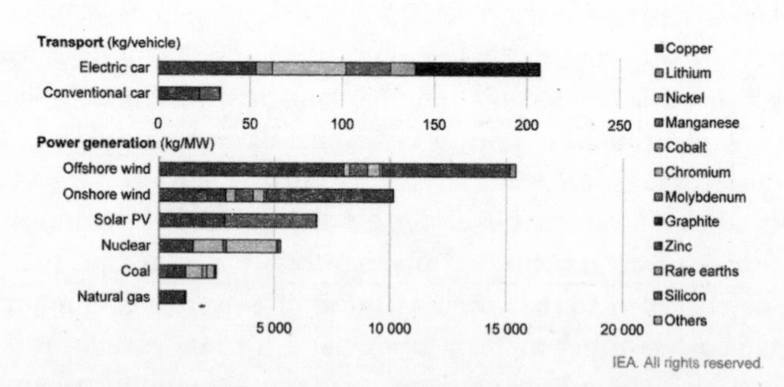

Figure 2.1 Mineral intensity of selected clean and fossil energy technologies

Source: IEA (2021) 'The role of critical minerals in clean energy transitions', World Outlook Special Report, https://iea.blob.core.windows.net/assets/ffd2a83b-8c30-4e9d-980a-52b6d9a86fdc/TheRoleofCriticalMineralsinCleanEnergyTransitions.pdf

nearly 35 per cent, silver by 38–105 per cent and indium by more than 300 per cent.

Consider that an electric vehicle can use up to six times more minerals than a conventional car, requiring up to half a million pounds of raw materials to be mined and processed. In July 2022, the IEA's 'Securing Clean Energy Technology Supply Chains' report raised the alarm on availability of critical semiconductors and minerals and made the point that 'we must ensure that the path out of the current energy security crisis and the race to net zero emissions do not simply replace one set of concerns with another'.[7] It pointed out that supply constraints had already applied the brakes to the acceleration of the electric vehicle market. 'In May 2022, Tesla and Volkswagen warned that supply chain disruptions and higher raw material prices threatened the rollout of EVs, with demand threatening to exceed production capacity. Companies are starting to scale

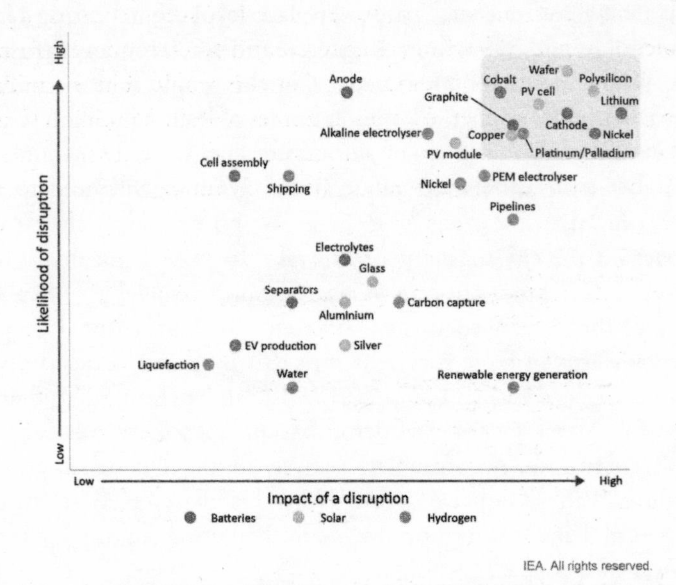

Figure 2.2 Likelihood and magnitude of the impact of potential supply disruptions for leading clean energy inputs

Source: International Energy Agency (2022) 'Securing Clean Energy Technology Supply Chains', July.

back their EV production targets, while higher prices could also delay the achievement of cost parity with conventional internal combustion engine vehicles. Volkswagen sold out of EVs in the USA and Europe in the first three months of 2022 due to shortages of semiconductors and wiring harnesses made in Ukraine. Tesla had agreed a deal with Piedmont Lithium for the supply of 53,000 tonnes of lithium per year for five years starting during the period July 2022 and July 2023, but shipments have been pushed back due to delays in obtaining mining permits. Nio, a Chinese EV manufacturer, had to suspend production in April 2022 due to local supply chain problems caused by COVID-19 restrictions and recently announced an increase in the price of its electric sports utility vehicle due to higher raw material costs.'

Demand for lithium (which needs 500,000 gallons of water to produce a single tonne, competing with local farmers) has been predicted to increase by as much as 2,700–4,300 per cent and cobalt and nickel by as much as 2,500 per cent. Global annual extraction of neodymium and dysprosium is projected to rise by 70 per cent and copper by more than 200 per cent. (Copper mining is more concentrated than oil production. 38 per cent of copper is mined in Chile and Peru.)

Beyond the immediate physical and governance challenges of extraction, there are other limits in terms of time, supply chains and impact on the environment. It can take 16 years from initial discovery to first production in mines. Given the challenges of sourcing and the time needed for extraction, the International Energy Agency warns of large shortfalls. But this is against a backdrop of a pre-existing over-extraction crisis. Mining has been identified as one of the largest drivers of deforestation, ecosystem collapse and biodiversity loss. According to analysis of the UN International Resource Panel Global Material Flows Database, even at current rates of global consumption, we are overshooting sustainable levels by over 80 per cent.

Limited supplies and increasing demand are driving up prices for the same scarce resources. After Russia's invasion of Ukraine, inflation rates in Europe and the USA soared to double digits driven by energy and other related factors. Higher rates of inflation together

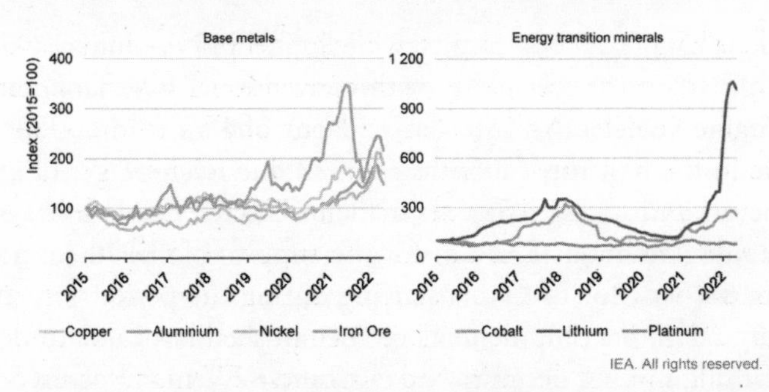

Figure 2.3 International prices of selected metals and critical minerals for clean energy technologies

Source: International Energy Agency (2022) 'Securing Clean Energy Technology Supply Chains', July.

with increasing interest rates puts stress on the economic feasibility of new infrastructure.

And then there is competition and the risk that it tips over the edge into conflict. Most of the key minerals needed for the energy transition are in the Global South. Parts of Latin America, Africa and Asia are at risk of becoming the target of a 'scramble for resources'. History echoes with precedents – gold and silver in Latin America in the seventeenth and eighteenth centuries, cotton and sugar in the Caribbean in the nineteenth century, diamonds from South Africa, cobalt from Congo and oil from the Middle East in the twentieth century. As Daniel Yergin points out: 'For many decades … oil and, at times, natural gas have been deeply enmeshed with geopolitics', but 'geopolitical clash has now become entangled with … new supply chains that are developing for the net zero carbon future.'[8]

The South China Sea

China needs energy to grow. The nexus to Russia is key. So is the 'Belt and Road' initiative – a global infrastructure-led economic development strategy involving over 140 countries, based on historical road, rail and maritime trade routes – and the South China Sea.

The South China Sea connects the Pacific and the Indian Oceans and control over it effectively determines control over global maritime trade. The South China Sea is one of the most important and valuable trading routes in the world, home to over US$5 trillion of trade every year, comprising more than 60 per cent of global maritime trade and more than 22 per cent of total global trade. It represents 65 per cent of China's total trade and over 40 per cent of Japan's, and 40 per cent of global petroleum products pass through it.

China has presented historical claims to 80 per cent of the South China Sea within a 'nine-dash line' as an extension of the continental shelf of China, despite the United Nations Convention on the Law of the Sea (UNCLOS), which limits the territorial boundaries of a country to 12 nautical miles and an area of contiguous waters to a further 12 nautical miles.

To reinforce its claims, China has now occupied the Spratly Islands, Paracel Islands, Mischief Reef, Subi Reef and Scarborough Shoal. The South China Sea is the location for military manoeuvres that many have referred to as 'an accident waiting to happen'. In 2021, the USS *Theodore Roosevelt* carrier strike force failed on a 'freedom of navigation operation' into a part of the South China Sea that China claims. Chinese bombers and jets responded with a 'mock attack', including simulating the launch of anti-ship missiles. In May 2022, a Royal Australian Air Force P-8 maritime surveillance aircraft was intercepted by a Chinese J-16 fighter aircraft, during what the Australian Prime Minister, Anthony Albanese, said was routine military surveillance activity. The Chinese jet released 'chaff' – an anti-radar device that includes small pieces of aluminium – that entered the plane's engine. Behind this dangerous manoeuvring is competition for resources, including those needed for the energy transition.

The South China Sea is rich in resources. It holds nearly one-third of the world's biodiversity, including the highest density of seafood and 15 per cent of the world's fisheries, thanks to its nutrient-rich tropical waters. In addition, there is an estimated 220 billion barrels of oil reserves, together with massive unexploited reserves of natural gas and rare earth metals.

Even closer to home for China, and sounding echoes of the Russian invasion of Ukraine, China warned the USA in June 2022 that any attempt to make Taiwan independent from China would trigger military action by Beijing's forces. China, said Defence Minister Wei Fenghe, would 'fight at any cost'.[9] Only in late May had Taiwan stated that it had deployed 22 fighter jets, early-warning and anti-submarine aircraft to warn off 30 warplanes sent by China into its defence zone. China and Taiwan were divided by civil war in the 1940s and Beijing has insisted that it will reclaim Taiwan in due course, by force if necessary.

Taiwan is the largest economy and the most populous state that is not in the United Nations. It has an export-driven economy, which is the seventh largest in Asia. It is a net food and energy exporter. Taiwan has rich coal deposits, some 180 million tonnes, most of it buried under the mountains in the north. Natural gas reserves were discovered in Guantian in 2012 and are estimated to have a production capacity of 1 billion cubic metres. The country has an estimated 100 tonnes of gold deposits. It has 30 million metric tonnes of marble, used in the construction and processing of products such as fertilizer, paper and carbide. It is one of the major exporters of cement, including to Malaysia, Ghana, Indonesia and Australia, and can produce 26 million tonnes of it a year. It also has abundant copper deposits, although all its mines are currently closed. Taiwan's main export products are electronics, basic metals and metal products, plastics and rubber, chemicals, and machinery. But amongst the most important of these are semiconductors – it now accounts for 92 per cent of the world's most advanced semiconductor manufacturing capacity.

China was outraged by a visit in August 2022 to Taiwan by Nancy Pelosi, Speaker of the US House of Representatives, during which she expressed support for the territory. (The Chinese government made a statement that it viewed the visit of Ms Pelosi and the accompanying US delegation as 'egregious provocations'.) China scrambled military forces to conduct exercises and war games, fired live ammunition and ballistic missiles over the island, closed air space, crossed the unofficial military buffer zone between China and Taiwan, sanctioned Ms Pelosi and announced that it was halting cooperation on climate change, military talks and other key matters. All in all, China believes, Taiwan's resources are worth fighting for.

Water Scarcity

In addition to 'energy security', 'mineral security' and 'battery security', we can add 'water scarcity' to the list of issues arising from resource constraints with the potential to spill over into conflict.

Following Russia's annexation of Crimea in 2014, Ukraine blocked the North Crimean Canal, which provided 85 per cent of Crimea's drinking and irrigation water. Ukraine's top irrigation official at the time, Vasily Stashuk, said that it would bring a humanitarian catastrophe amongst a population of 2.3 million, and the blockade nearly eliminated agriculture in Crimea. The problem was aggravated by a major environmental hazard in the northern Crimean town of Armyansk, where a major producer of titanium dioxide, chemical fertilizers, sodium and lithium had reportedly been pumping water from underground sources and filling aquifers with salty water, turning parts of northern Crimea into salty marshes. Fast-forward to 2021 and the already war-torn Donbas region was feared to be on the verge of environmental disaster as neglected and abandoned mines were filling with toxic groundwater, threatening to contaminate drinking water from rivers and wells in the area, as well as surrounding soil, making the land unfit for farming. Methane gas from the mines was also pushed to the surface, threatening earthquakes and explosions.

Water security is both global and local and is exacerbated by climate change. Over 70 per cent of natural disasters in the first two decades of the twentieth century were water-related, including droughts and floods. Extreme weather events make it more difficult to access safe drinking water and climate change exacerbates water stress from limited water resources, leading to increased competition for water and even conflict. For example, the war in Darfur, Sudan, is often cited as a 'climate conflict', within the context of an imbalanced political economy, severe racial tensions, and a breakdown in land management institutions. Local conflicts over resources have been traced back to the Sahelian droughts of the 1970s and 1980s.[9] Disputes over scarce water and grazing land have been attributed as key tiggers for conflict in Darfur.

The problem starts with the fact that 97 per cent of the Earth's water is saltwater and only 3 per cent is freshwater. Only 1 per cent is readily available for use. Competition for its use is increasing from rising population, power, irrigation, agriculture and industry. Demand is set to outstrip supply. The United Nations has warned that the planet has lost more than half of its natural wetlands in the last 100 years. McKinsey predicts that by 2030, over half of the world's population will live in water-scarce areas and that 40 per cent more water will be needed to meet global demand. Managing, re-using, exchanging, smart metering and improving efficiency (like energy as we will cover later) and restoring natural ecosystems could reduce consumption and emissions by 20–25 per cent. The OECD estimates that over US$22 trillion of cumulative investment in water infrastructure will be needed by 2050.

Food Security

Then we come to 'food security'. Along with energy, Russia's invasion of Ukraine has highlighted the world's dependence on Ukraine and Russia for food. (Russia's wheat exports doubled between 2015 and 2020, making it the largest wheat exporter in the world, supplying around a quarter of the global market.) Europe sources most of its wheat and sunflower oil from Ukraine and Russia. Interruption of supplies from the blockading of the Black Sea export routes wasted stockpiles of food and raised fears of global food supply shortages, price spikes and hunger.

In May 2022, the Russian blockade of Ukraine's Black Sea ports presented the prospect of a global food crisis. Russia was accused of 'weaponizing' energy by shutting off natural gas to Germany (through the Yamal pipeline) and Finland, and the supply of electricity to Sweden. By 18 May 2022, both Sweden and Finland had formally applied to join NATO. Russia was also accused of weaponizing food, a limited resource, and another issue entering the global geopolitical stage. The United Nations estimated that nearly 25 million tonnes of grain were stuck in Ukraine under blockage on 6 May 2022, supporting fears of food shortages in dozens of countries.

Ukraine provided 42 per cent of the world's sunflower oil exports in 2019, 16 per cent of corn exports, 10 per cent of barley exports and 9 per cent of wheat exports.

Food security interacts with energy and the climate. Record temperatures in India in May 2022 forced the country to declare a suspension of wheat exports because of crop failure. Widespread food shortages were widely predicted to lead to continued widespread increases in food prices and potential humanitarian disaster. According to the World Bank, a 1 per cent increase in food prices forces 10 million people into extreme poverty.[10]

A pinch point for Russian and Ukrainian wheat exports is North Africa and the Middle East, regions, like India, that are on the front line of the climate-related temperature rises. The blockage of wheat from Ukraine, together with the increase in the price of the average food basket by up to 350 per cent (partly driven by higher fertilizer, energy and transport costs), sparked fears of the worst food security crisis in 50 years, with the potential, according to the World Food Programme, to plunge 50 million people into starvation.[11] Egypt, where the word for 'life' – *aish* – is also used to refer to food, and which hosted the COP27 United Nations climate change conference in 2022, imports almost four-fifths of its wheat from Ukraine and Russia. Memories of revolutions, which toppled regimes across the region a decade before, remain fresh, haunted by the chants for 'bread, freedom, social justice' on the streets of Egypt during the Arab Spring.

Efficiency and Waste

As we will cover later, some two-thirds of energy is lost before it reaches the point of use, according to the World Economic Forum. Roughly one-third of the food produced in the world for human consumption gets lost or wasted each year, according to the United Nations Environment Programme.[12] This is equivalent to 1.3 billion tonnes, worth roughly US$1 trillion, and is responsible for 8 per cent of global greenhouse gas emissions, according to the United Nations Food and Agriculture Organization.[13] Over 300 billion litres of water are lost every year from leakage and waste – some 20–40 per cent of Europe's available water is estimated to be wasted – much of which relates to consumer food waste. Chronic resource waste starts at the very beginning of the food production line. Around 60 per cent of nitrogen poured onto land to fertilize fields is lost. The Haber-Bosch process behind ammonia-based nitrogen fertilizer itself consumes 3–5 per cent of the world's natural gas production and accounts for roughly 1 per cent of global annual greenhouse gas emissions, more than any other industrial chemical-making reaction, while nitrogen use efficiency is typically less than 50 per cent. Much of the nitrate from fertilizer runs off into soils, rivers, and oceans, creating 'dead zones'. If not applied properly, up to 40 per cent of urea fertilizer can escape into the atmosphere as ammonia gas, through a process called volatilization. Nitrous oxide is now the third most important greenhouse gas after carbon dioxide and methane. While scientists are working to reduce how much greenhouse gas the ammonia-making process emits, solutions to these problems cannot be addressed on the supply side alone. We must be more efficient.

The issue of resource efficiency applies to almost all areas of the economy and is often most stark in those areas that collide most seriously with the environment. Take plastic packaging, the disposal of which is leading to severe ecological damage to land, rivers and seas. A 2016 study by McKinsey, the World Economic Forum, and the Ellen MacArthur Foundation found that, of the annual 78 million tonnes of plastic packaging produced, 40 per cent ended up in landfill, 32 per cent leaked into the environment (including an estimated 8 million tonnes into oceans) and 14 per cent was incinerated. Only 14 per cent was collected for recycling.[14]

Similarly, according to the Global Fashion Agenda, less than 1 per cent of textile waste is recycled into fibres for new clothes. While clothing production doubled between 2000 and 2015, the use of an item of clothing decreased by 36 per cent, according to the Ellen MacArthur Foundation.[15] All this matters because globally the fashion industry was responsible for around 4 per cent of total greenhouse gas emissions in 2018, according to McKinsey. Fossil fuel-based synthetic materials and recycled synthetics make up more than half of total fibre production and according to the European

Commission, up to 500,000 tonnes of synthetic fibres from textiles are released into oceans every year.

Even if it is collected, it doesn't always get recycled. Bloomberg News reported in 2022 that soft plastics collected at a major UK supermarket chain travelled more than 1,000 miles from the UK to Poland, only to end up being sent to landfills or burned.[16] Plastics that can be recycled need around a fifth as much energy as making new plastic. As plastics are produced from petrochemicals derived from crude oil, the same stuff that nations are fighting for, we have created a perverse and vicious form of circular economy. The circle must turn the other way.

The concept of a circular economy has become fashionable in sustainability circles but eludes the fashion industry itself. Apparel and footwear together contribute some 8 per cent of global greenhouse gas emissions, about the same as beef and pork. The fashion industry is powered by coal, oil, diesel, jet fuel and bunker crude for ships. Waste from production goes directly to land and water. Twenty-one trillion gallons of water a year are used, and 203 trillion pounds of waste are produced, while 30 per cent of produced clothing (like food) never makes it to the customer. Between 1992 and 1994, at the University of British Columbia in Vancouver, William Rees and Mathis Wackernagel developed the concept of the ecological footprint (of which the narrower carbon footprint is a component – the carbon footprint is the fastest growing part of the ecological footprint and is currently estimated to represent around 60 per cent of the total) and a calculation methodology.[17] The model quantifies the amount of environmental resources that it takes to produce goods and services. It enables a comparison of consumption on the one hand and biocapacity on the other hand. Since 2003, the Global Footprint Network has been using UN data sources to calculate the ecological footprint of the world as a whole and over 200 countries. In 2019, it estimated humanity's ecological footprint as 1.75 planet Earths, or in other words, that humanity's demands were 1.75 times more than what the planet's ecosystems renewed. (The average biologically productive area per person worldwide was estimated, based on 2018 data, to be 1.6 global hectares (gha) per person. The ecological footprint per person was estimated to be 2.8gha.)[18]

Scarce and limited resources are being consumed at an unsustainable rate, with both geopolitical and global ecological consequences. A supply-driven economy feeding growing consumption will inevitably incentivize production over efficiency. But unless we reverse this, the results are environmental degradation, climate change and conflict.

- Mineral security hit the headlines more regularly after the LME's nickel crisis and revealed the intense global competition for 'rare earths' and other key resources, many of which are vital for the clean energy transition.
- Consumption of minerals exceeds sustainable supply and the clean energy transition will make this even more challenging.
- Access to resources is a defining feature of the politics of the South China Sea and between China and Taiwan.
- Meanwhile, we are wasting most of the water, food and energy that we are fighting over.

CHAPTER 3

Climate Change and How All Need Not Be Lost

• • •

- How a glimpse of the limits changed everything for me.
- The limits of energy policy to date.
- What is happening to our climate and why it matters.
- Whether and how we can turn it around and how much it will cost or benefit us.
- The roles and limits of electricity and gas.
- Why more (of the same) is less (effective) and why less is more.

I was at London City Airport in February 2006. I had managed to arrive keenly early to take a flight for a break in the mountains in Switzerland, but I had left my passport behind. I had sent for it but had to wait for it to arrive. While I waited, I went to the bookstall to look around. I bought a copy of the latest edition of Jared Diamond's book, *Collapse: How Societies Choose to Fail or Succeed*. The premise was compelling: how environmental problems contribute to the success or failure, or at extremis the demise, of societies. It was multi-disciplinary in that its 'input variables' weaved together geography, political science, economics and history. The premise was that, after looking back at societies in history that didn't make it, like the Greenland Norse, the Anasazi, the Maya and the inhabitants of Easter Island (with the haunting image of the person that cut down the last palm tree), there were five factors that contributed to collapse:

climate change, hostile neighbours, loss of essential trading partners, environmental problems and how societies respond to each of these challenges. The pivot point of collapse was 'the limit where impact outstrips resources'.[1] It concluded with an outlook on China, which would be my next destination for business after my winter break. It was more than a good read; it was a contemporary seeing his place in the patterns of world history. It changed the way I looked at, and thought about, everything.

As soon as I got back to work, I was on a flight to China. 2006 was an important year in an important period in China. It was as much a turning circle as a turning point. China had been a member of the World Trade Organization for five years. Massive urbanization was under way, bringing with it huge population movements and bringing hundreds of millions of people out of poverty. China had just launched 'a grand experiment' with the National Medium- and Long-Term Plan for the Development of Science and Technology. Concerns about energy security were absorbed into an environmentalist world view in China's 11th Five-Year Plan in 2006, focused on economizing on resources, a 'cycling economy' and an 'environmentally friendly society', with local officials instructed to 'rethink, reduce, re-use, recycle and repair'. An incredibly dynamic market economy was at work, within the context and confines of state command. The people, public places and atmosphere in Beijing were astonishing and exciting. Shanghai was an electric fusion of the past and the future. Standing on the roof of a building in the Bund near the French Concession, a bright red communist flag was illuminated by searchlight, overlooking the space-age Oriental Pearl Tower, at the time the tallest building in China. But the environmental problems were visible with the naked eye. If I couldn't see down the street for pollution, I could see why by understanding the patterns.

Back in the West, former US Vice President Al Gore's *An Inconvenient Truth* was released later that year, and then in 2007, the United Nations' Intergovernmental Panel on Climate Change (IPCC) Fourth Assessment Report (AR4) was published. 'Warming of the climate system is unequivocal', it said, and it was 'very likely due to the observed increase in anthropogenic GHG concentrations'.[2] The Nobel Peace Prize in 2007 was awarded jointly to the IPCC and

Al Gore 'for their efforts to build up and disseminate greater knowledge about man made-climate change'.[3] In 2007, a new act in the story was starting and I joined in.

Today, scientists have reached a broad consensus that climate change is occurring, and that human pollution is contributing to it. A range of consequences of climate change, from the bad to the dire, has been established. But if scientists now agree at least as to the root

Global greenhouse gas emissions by sector

This is shown for the year 2016 – global greenhouse gas emissions were 49.4 billion tonnes CO_2eq.

Our World in Data

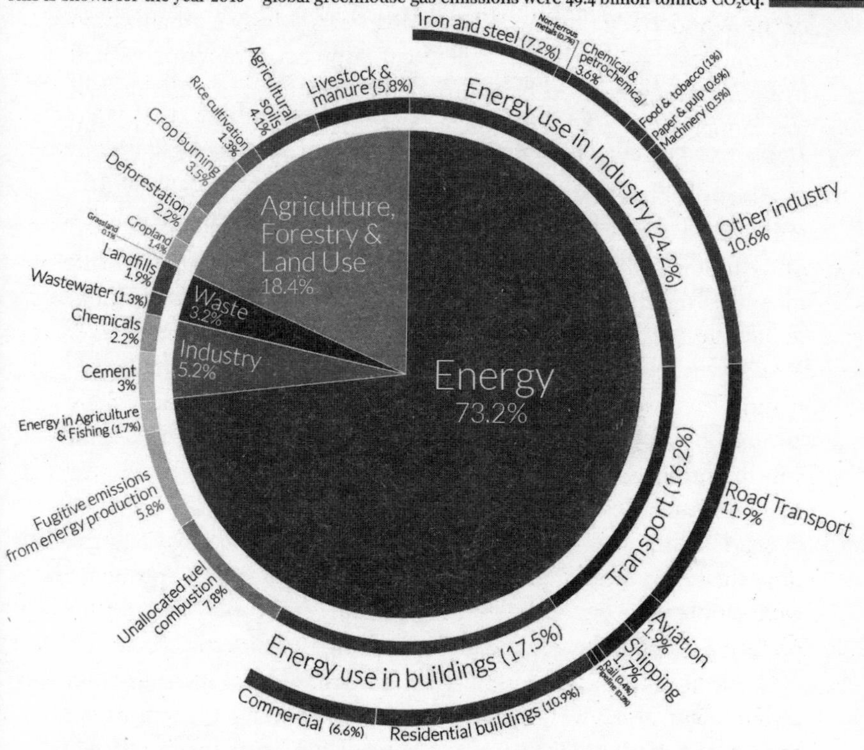

OurWorldinData.org – Research and data to make progress against the world's largest problems.

Figure 3.1 Global greenhouse gas emissions by sector

Source: Our World in Data, 'Emissions by sector', https://ourworldindata.org/emissions-by-sector licenced under the CC https://creativecommons.org/licenses/by/4.0/

causes and potential scenarios, there is still no such political agreement as to what to do about it.

The story of climate change often starts with energy as the lead protagonist because it is associated with most of the human-made greenhouse gas emissions that contribute to it. Very broadly, around 80 per cent of the story relates to energy and around 20 per cent to land management (see Figures 3.1 and 3.2).

The Limits of Policy Responses to Date

Today's conflict or tension is that it may also appear that policies applied to date to mitigate climate change by abating emissions have had no tangible effect on overall emissions, which have gone up, not down. Have we therefore reached another limit, that is the impact of current climate and energy policy? So far, substantially all mitigation policy and efforts and thereby investment of trillions of dollars has gone towards lower-carbon energy generation such as renewables. This looks set to continue, with massive public and private sector financial and moral support. Renewable energy is widely considered the largest and most important antidote to carbon emissions and policy responses, for the most part, involve

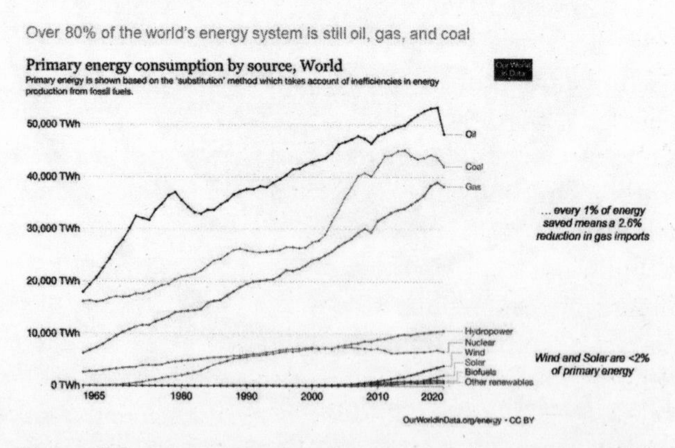

Figure 3.2 Primary energy consumption by source, world

Source: Our World in Data, based on BP Statistical Review of World Energy https://ourworldindata.org/grapher/global-energy-consumption-source licenced under the CC

market incentive-based hammers to, following the analogy, hit various types of renewable energy nails. However, energy generation and carbon emissions have continued to rise and so, consequently, has the concentration of greenhouse gas in the atmosphere. There is no evidence that this is set to change anything like enough on the current trajectory.

Current policy is having only a marginal impact and has not resulted in a reduction of overall emissions. We would have marginally more carbon emissions had renewable generation not been deployed at a substantial scale in many countries and the prospects for renewable power would be lower had we not seen costs fall as capacity increased, the 'experience curve'. However, renewable power is not yet contributing more than a minority of energy supply, as 80 per cent of the world's energy system still consists of oil, gas and coal. (The world burns approximately 100 million barrels of oil a day, 23 million tonnes of coal and 10 billion cubic metres of natural gas, which collectively emit around 33 billion tonnes of CO_2 every year.) Renewable power, which can be intermittent (in other words, available only when, for example, the sun shines and the wind blows), often requires conventional energy and storage to balance the grid and to ensure that energy is available when it is needed. Emissions would also have been higher had substantial improvements in energy efficiency and energy intensity not been made in many countries. However, this too has been marginal and most of the opportunity remains to be taken in an energy system where, in many countries, most energy is lost or wasted.

While current climate policies tend to be aimed at achieving net zero carbon emissions by or soon after the middle of the century, the achievement to date has been net zero carbon emission reductions. Emissions have risen by, on average, two parts per million per annum, and they continue to do so. One of the reasons that we are, in fact, going backwards on emissions is that we are producing, adding, more energy and greenhouse gas emissions, not less, and this is set to continue. Renewable energy generation at massive scale takes time and can be expensive to build, needs balancing and depends on scarce resources. Meanwhile we are not displacing conventional energy, but we are adding to it. At the same time as

Most of the world is still the wrong shade of (pale) green ...

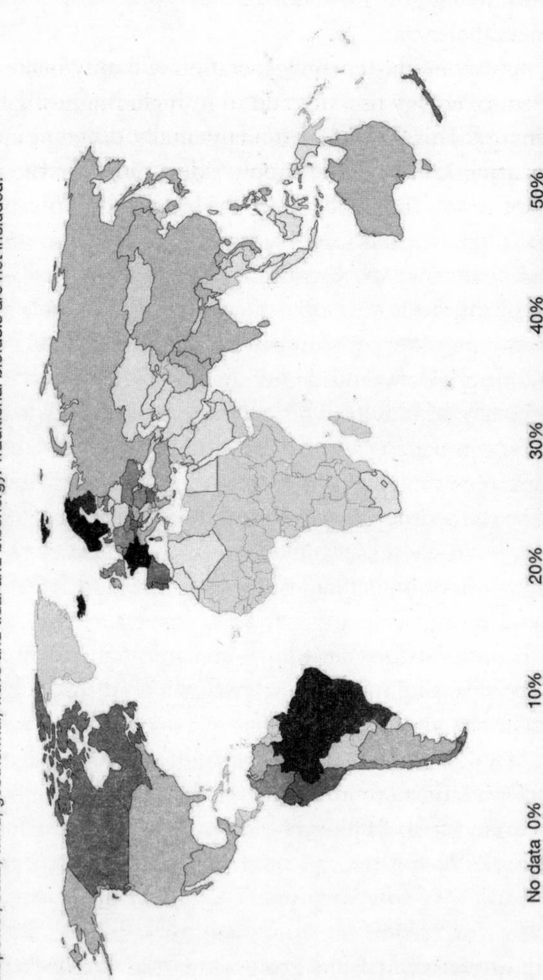

Figure 3.3 Share of primary energy from low-carbon sources, 2020

Source: Our World in Data, based on BP Statistical Review of World Energy, https://ourworldindata.org/grapher/low-carbon-share-energy licenced under the CC

we invest in renewable energy, natural gas use is set to rise by over 50 per cent by 2050, the date that most countries are targeting for net zero. More coal capacity is being added in China every year than the total energy capacity of the UK, by a factor of nearly three. And consumption is going up, not down, a great deal of which is a function of inefficiency.

Adding new renewable energy generation will only make a difference if we reduce energy use, not add to it, including using less conventional energy. This requires a fundamentally different approach, shifting our attention from the supply side to include the demand side of the equation. The opportunity for improvement is massive as most energy in the world is lost or wasted, somewhere in the process of generation, transmission, distribution, and end use. Efficiency, the demand side, is literally the other side of the coin. One side of the coin, on the supply side, costs money. And it can be slow. The other side of the coin, the demand side, entails using less, which costs less, which improves productivity, which is more profitable, which makes money. And it can be fast.

Producing renewable energy at massive scale will take time and require cost competitiveness versus conventional energy. We can hope and expect to see increasing levels of penetration of renewables in power and heat production, ramping up towards the 2030s and 2040s. However, displacement of conventional energy will be needed in the meantime to reduce emissions, together with a minimization of energy lost or wasted to reduce consumption, or to put it another way, to reduce the amount of new energy that needs to be used or added. While renewable energy may offer a transformation in carbon intensity in the medium to long term, major effective solutions need to be delivered in the short term. That is partly because we now know that we probably do not have the luxury of the long term to change the game.

The limits for carbon emissions are approaching fast. Unless human carbon emissions from energy and land use peak by 2025, then, it is predicted, we will not be able to limit temperature rise to levels that would avoid the worst, certainly disruptive and dislocating, potentially catastrophic, effects of global warming.

What Is Happening?

The Intergovernmental Panel on Climate Change is a Nobel Prize winning body of the United Nations. It was created in 1988 (a year after the definition of sustainable development by the World Commission on Environment and Development) by the United Nations Environment Programme and the World Meteorological Institute and comprises 195 member countries. It provides the scientific and academic cornerstone for research, scenarios, and projections. The IPCC, together with a broad global consensus of scientists, now suggests that we are reaching a new edge, a limit, a turning point.

Increased levels of carbon emissions from energy and land use are now proven to be generating more greenhouse gases that have a warming effect on the Earth's atmosphere than are being removed by the Earth's natural capacity to absorb them through oceans and vegetation (the 'carbon cycle'). Carbon dioxide is the largest and most persistent greenhouse gas. Higher carbon dioxide levels in the atmosphere increase temperatures. The National Oceanic and Atmospheric Administration's (NOAA) monitoring station at Mauna Loa, Hawaii, averaged a reading of 421 parts per million of CO_2 for the month of May 2022, a milestone of 50 per cent higher than pre-industrial times. CO_2 levels were 280 parts per million before the Industrial Revolution in the late nineteenth century. Scientists have been hoping not to exceed a limit of 350 parts per million. Current levels of atmospheric CO_2 are approximately the same as 4.1 to 4.5 million years ago in the Pliocene era. Temperatures were 3.9°C hotter and sea levels were 5 to 25 metres higher. South Florida (which suffered the deadly Hurricane Ian in 2022) was under water. Human civilization, as *Time* magazine pointed out, has never known these conditions. Temperatures have already risen and are on track to rise to dangerous levels.[4] Therefore, emissions need to be reduced to avoid the worst effects of breaching the limits and stability of the carbon cycle.

The emissions limits have been defined in terms of a carbon budget and the IPCC holds the pen as a source of what they are likely to be for any implied temperature rise. The limits to stay within a 1.5 to 2.0°C implied temperature rise are likely to be somewhere between 500 and 1000 gigatonnes of carbon and at the current rate

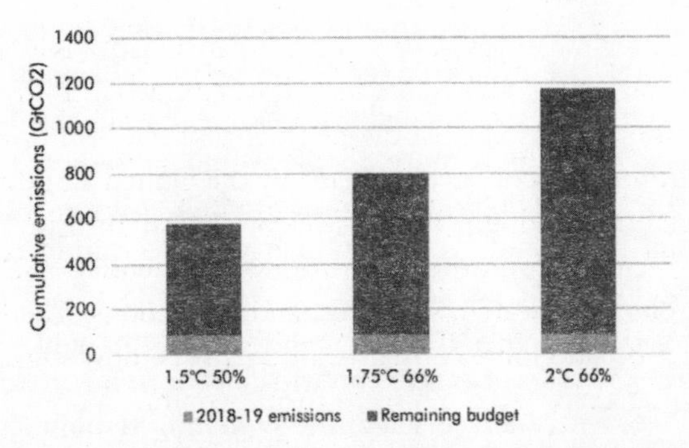

Figure 3.4 We have only 8–10 years left of carbon budget to limit global warming to 1.5°C

Source: Carbon Tracker, 'Carbon budgets: where are we now?', https://carbontracker.org/carbon-budgets-where-are-we-now/

of emissions (which is in fact rising not falling), the budget is due to be spent completely within the next 10 years. To add to the problem, as it worsens, we may lose control. For instance, as the Earth warms, so permafrost thaws, releasing carbon dioxide and the even more potent, albeit less persistent, methane. This could represent as much as one-third of the whole carbon budget.

A number of scenarios have been produced by the IPCC. The median outcome from all scientific scenarios deemed plausible is estimated to imply a global temperature rise of approximately 2.2°C, higher than the current global UN target of 1.5°C, although some modelled scenarios for unabated climate change demonstrate a risk that temperatures could rise 4.5–6°C. At higher levels, all effects are worse: ice caps melt faster, accelerating carbon emissions, oceans acidify faster, reducing their effectiveness as carbon sinks, global climate patterns keeping the Earth's biosphere in balance are broken and areas on Earth become uninhabitable. As the planet warms, we face further extinction of animal species, more water shortages, more displacement, and conflict of people unable to live in overheated conditions, severe weather events including droughts and wildfires, and sea level rises at a level that threatens flooding of many of the world's great cities. However, higher levels of temperature rise are not needed for society to be damaged by the impact. Even if the global average surface temperature of the Earth rises more than 1.5°C to 2.0°C from pre-industrial levels (below the median of plausible scenarios), some of the worst effects of climate change would be felt (for instance, 1.7 billion more people would be subject to severe heatwaves, sea levels could rise 10 centimetres and we could lose almost all coral reefs) and human influence over limiting climate change may no longer be possible.

The IPCC's standards for establishing cause and effect are high, requiring 30 years of data to confirm detection and causal link. Already, the IPCC has observed scientifically demonstrable increases in heatwaves, droughts, extreme precipitation and 'fire weather'. Europeans suffered record temperatures in the summer of 2022. This was attributed to hot air being carried from Africa via a 'double jet' pattern in which the jet stream branches in two as it crosses Europe. I wrote a significant section of this book in the Douro Valley upstream from Porto. We managed to take a family rabelo boat trip to the exact place on the exact day as the hottest July day ever recorded in Portugal, at 47.3°C. We lived to tell the tale. Back at our hotel, although slightly cooler, it was raining ash from the nearby forest fires.

Decisions we make now lock in future carbon emissions and, based on the science, make climate change and the danger it poses to society worse. From where we stand today, global temperatures have already risen over 1.1°C since pre-industrial times. Arctic sea ice has receded 13 per cent since 1979, we have lost over 420 billion metric tonnes of ice sheets per year since 2002. Sea levels have risen 10 centimetres since 1992 and oceans have heated by more than 330 zettajoules since 1955. Glacial ice melt has caused the Earth's axis to shift because of the redistribution of the planet's water. Record temperatures are being recorded worldwide, leading to heatwaves, droughts, power failures, wildfires, hospitalizations, displacement and deaths. The World Meteorological Organization declared the decade 2010–20 the warmest since modern records began in the 1880s. For 98 per cent of the Earth, the twentieth century was warmer than any time in the last 2,000 years.[5] The world is now warmer than at any time in the last 125,000 years. And it looks set to get warmer.

Greece, one of the hotspots in the eastern Mediterranean, appointed its first 'climate crisis' minister in 2021 following huge wildfires close to Athens that summer, consisting of 65 simultaneous forest fires a day by late August, that destroyed 1.3 million hectares of woodland. California's drought conditions in the first half of 2022 brought back extreme water shortages, presaged wildfires and threatened to reduce the state's hydroelectric generation by almost 50 per cent in the summer, compared to 'normal' water conditions. Salt Lake City faced the prospect of being a city without a lake as Great Salt Lake's surface area had shrunk by over two-thirds to the lowest levels on record, threatening drinking water. (Republican state lawmaker and rancher, Joel Ferry, described the situation as equivalent to an 'environmental nuclear bomb' for Utah, as saving the lake would mean competing for water with residents and farmers, whereas failing to act not only threatened animal life but human life, with the risk that high levels of arsenic would be exposed to winds, which could poison up to three-quarters of Utah's residents.[6]) Snowmelt that normally travels hundreds of kilometres from the top of mountains to taps, sustaining the growing populations of Las Vegas, Phoenix, Salt Lake City, Los Angeles and San Francisco, might no longer be relied on. In Europe and Asia, several countries were again announcing record temperatures in April and May 2023, including Vietnam which reached its highest ever temperature at over 44°C.

Global climate change creates extremes of cold as well as heat. In the same week as heat-induced power outages took down power supplies in Michigan, Missouri, Ohio, Wisconsin and Illinois, parts of Australia suffered a cold snap bringing snowfalls to southern states, increasing demand for heating, straining the grid and prompting New South Wales to invoke emergency powers requiring miners to redirect any coal that was meant to be sent overseas to local generators.

Back in Europe, summer heatwaves started early in 2022. Low levels of the Po River in Italy and the drought that followed led to the declaration of a state of emergency due to water shortages. France recorded its hottest May on record when temperatures reached 39°C in the Rhone valley. By August, French Prime Minister Elisabeth Borne activated a special crisis unit to tackle what she said was the worst drought in the country's history. All but three of the country's 96 departments had water restrictions in place and two-thirds were classified by the environment ministry as in 'crisis'. In the UK, amidst record temperatures and wildfires, the source of the River Thames dried up and was now more than five miles downstream, foreboding competition for scarce resource and loss of aquatic life. Fires had already started in Spain and Germany in June amidst heatwaves where temperatures reached 43°C and 39.2°C respectively. Temperatures had hit a record 47°C in Portugal in July with up to 170 forest fires reported a day. The European Forest Fire Information System reported the prevailing burnt area as four times greater than 2006–21 averages. The European Commission declared the summer droughts as the worst in 500 years. In California, the scale of the Oak Fire close to Yosemite National Park marked a dramatic start to the wildfire season, exacerbated by drought, overgrown vegetation and climate change. The Met Office weather agency declared 2022 the warmest year since records began in 1884. France, Switzerland and Spain also reported the same phenomenon. Germany, Belarus, Belgium, Latvia, Poland and the Netherlands all had record daily highs for the New Year.

As we came into 2023, new records were set for all-time-high temperatures during April in Asia, Europe and South America.[7] In the UK, June 2023 was the hottest since records began almost 140 years ago. The EU-funded Copernicus Climate Change Service (CS3) reported that average global surface temperatures were more than 1.5°C above pre-industrial levels for several days, the first time this had happened in the northern hemisphere summer. Global average sea surface temperatures had already broken records in April and May (attributed to a concert of human-made global warming, the natural warming phase of the El Niño Southern Oscillation, Saharan dust patterns and low-sulphur shipping fuels). Areas of North America were reported to have been 10°C above the seasonal average in June. The same month, records were broken in Beijing, China, and countries as far apart as India, Spain, Vietnam and Iran sweltered in extreme heat. Parts of Canada and the East Coast of the United States, including New York City, choked on the hazardous haze of forest fires burning in Canada and emitting an estimated 160 million tonnes of carbon. Then, on Monday 3 July, the US National Centers for Environmental Prediction stated that the world's average temperature had reached an all-time high of over 17.01°C. New high temperature records would be set around the world in the often deadly 'heat storms' that ushered in the summer of 2023.

Turning It Around

But all need not be lost. A significant change in scientific consensus was signalled when in December 2020, Dr Joeri Rogelj of London's Grantham Institute and a lead author of the Sixth Assessment of the IPCC stated: 'It is our best understanding that, if we bring carbon dioxide [emissions] down to net zero, the warming will level off. The climate will stabilize within a decade or two. There will be very little to no additional warming. Our best estimate is zero.'[8] Fundamentally, the message is that climate science now indicates that we are not necessarily locked into centuries of global warming, but that we have an opportunity to stop it after we achieve zero, or net zero (a balance between greenhouse gases put into the atmosphere and those taken out), carbon emissions. Net zero is therefore a threshold beyond which it is possible to reduce atmospheric CO_2 levels back to pre-industrial levels. It is a milestone which could bring the world back from the edge.

Whether or not, and if so to what extent, we have locked in climate change given emissions to date, either way, time is of the essence. To solve the problem, or at least to limit implied temperature rise from tipping over the edge, according to the IPCC, emissions need to fall by at least 8 per cent per annum for the next 20 years, maximizing the life of our carbon budget. Every year that emissions go up rather than down, the task becomes larger and compounds, as does the size of the action that we need to take in ever shorter periods of time. To live within the 1.5°C limit, it is understood that we will need to reduce emissions by 50 per cent every 10 years from now. The longer we wait to act, the more carbon is emitted and is added to warming atmospheric stocks, the more that the carbon budget is depleted, and so the harder the task in a shorter timeframe. If we fail to act before 2025, emissions reductions will need to be even more severe as the challenge gets bigger and the time to deliver it even shorter. By some measures, if

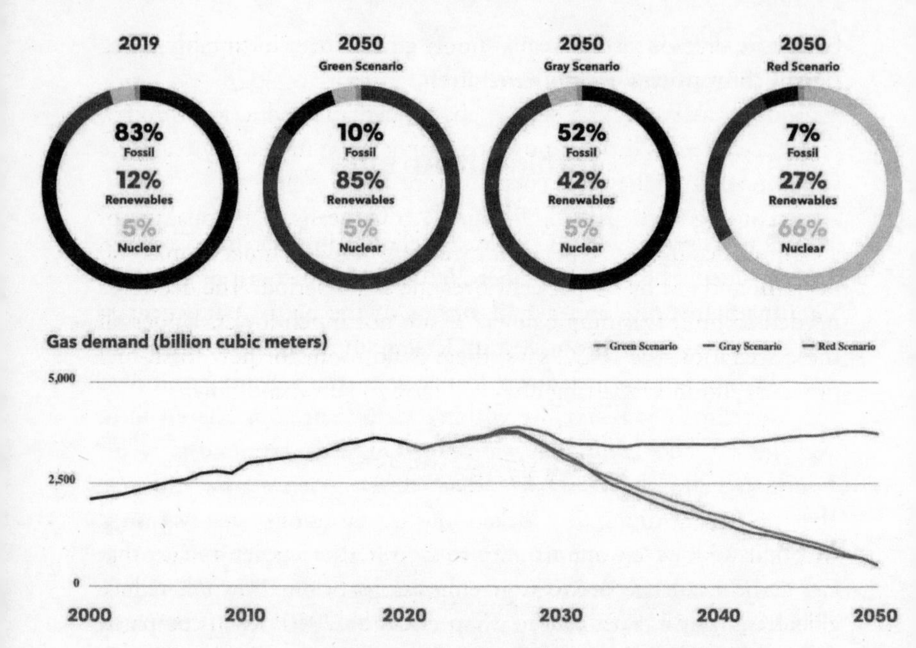

Figure 3.5 Low-carbon pathways and their potential impact

Source: Bloomberg NEF 2021 New Energy Outlook, https://about.bnef.com/new-energy-outlook/

no effective action is taken that reduces emissions and they stay at pre-pandemic levels, then we will have used up our carbon budget by the end of the decade and hit our limits.[9]

The largest-scale actions that we can take over time are widely understood to include shifting from fossil fuels to renewable energy generation, removing carbon by adding forest cover, capturing and using carbon, and researching and developing new technology. Many of these are solutions for the decades to come once the regulatory, physical, and financial demands have been overcome. Urgent action that is capable of implementation at the scale required above will need almost immediate step changes in demand as well as supply. That is, efficiency. Efficiency – delivering reductions in carbon emissions by reducing energy used by cutting energy wasted – has its greatest role to play this decade. If we get it right now, then less of the renewable energy that is generated when it becomes available will be lost or, like the last 30 years, simply added to an inexorably rising tide of conventional energy generation.[10]

Indeed, as Figure 3.5 shows, absent demand reduction through efficiency, there is little or no impact on gas use under any plausible scenario adding renewable energy before 2030.

According to the IPCC, pursuing 1.5°C means that global use of coal must decline by 95 per cent by 2050.[11] Oil use must drop by 60 per cent and gas by 45 per cent over the same period. The decreases needed to limit warming below 2°C are not much lower. Under all these scenarios, the IPCC says, there is no room for new fossil-fuel projects, and most existing ones will have to be wound down.

At What Cost?

All of this appears to come at a substantial cost. The report estimates that taking the actions needed to keep temperatures below 2°C could reduce global GDP by 1.3 per cent to 2.7 per cent by 2050, when compared with sticking only to the climate policies that countries have already announced.[12] However, the report points out that these costs do not account for damage caused by climate change, or the sacrifices required for adaptation. In other words, not adopting aggressive climate policies brings its own costs. These include lost lives, livelihoods, destruction caused by extreme weather events and lost productivity, to list but a few.

The costs of inaction depend to some extent on whether and how you value the future and how much more value you attribute to the present than the future. Either way, the numbers are stark. The Stern Review on the Economics of Climate Change in 2006 remains one of the largest, most important and influential reports of its kind. It pointed out that the costs of action now far outweigh the costs of inaction, which could be equivalent to a loss of 5 per cent of global GDP each year. While it places almost as much value on the future as the present, which is not how financial markets tend to produce valuations, even economists applying a lower value to the future than the present calculate a very substantial loss from inaction.

There are limits to the arguments that we should spend now to save later, or at least limits to what financial markets and business can do without financial or commercial rationale. These limits are often addressed through government-backed market incentives or payments, which create revenue streams now to recognize benefits in future. Examples include renewable energy feed-in tariffs, contracts for difference and the issuance of green certificates, often funded by green taxes, receipts from the issuance of certificated carbon emission allowances and other sources. These limits are tested by the political will of society to pay for market incentives through higher bills or taxes. These limits are exacerbated when they collide with national or commercial competitiveness or productivity, let alone the financial returns demanded by the capital markets. One of Europe's leaders in renewable energy incentives and deployment, Germany, has spent particularly heavily on downstream renewable energy generation and, in the process, exported much of this value by funding the revolutionary development of the upstream supply chain in China. Sadly, energy supply constraints today are forcing Germany back to coal in the short run after a period of subsidizing Russian gas.

This background is an important part of the rationale for why efficiency, which offers improvements and competitiveness, matters so much. Energy efficiency is the largest, cheapest, and fastest source of emissions reductions – and remains largely untapped. One of the most important and extraordinary features of energy efficiency is that it does not cost money but saves it. Reducing energy use cuts costs as well as carbon. As the IEA points out, every US$1 invested in energy efficiency results in US$2 of savings.[13] Energy efficiency is not only the largest but also the cheapest source of greenhouse gas emissions. It is a massive source of potential economic productivity and growth. If we are wasting two-thirds of one of the largest and most valuable inputs into the global economy, no wonder global economic productivity has been languishing.

The Electric Shock

An important war cry of the decarbonization movement has been 'electrify everything'. This has its limits. The idea is that we switch from energy from fossil fuels to renewably sourced electricity and shift to a new paradigm for carbon intensity. There are two fundamental problems with the argument if we are to limit temperature rise fast enough.

The first is that by electrifying everything, which is at least currently impossible, we would massively increase the amount of electricity that is needed to power the economy, by some estimates up to four times current levels. To give an idea of the scale of the problem, recall that only around 20 per cent of today's global energy system is electricity (the remainder comprises oil and petroleum products, including for transport, and natural gas and solid fuels, including for heat) and that only a minority is renewable. We would first have to green all electricity before we electrify all other parts of the energy system, such as heat and transport. So far, it is estimated that renewables have decarbonized around 10–15 per cent of industrial energy use and that the global fleet of electric cars, vans, trucks, buses, and bikes reduces global oil consumption by some 2 per cent.[14] There is a long road ahead.

To give an idea of the scale of the problems, let's start with the 'easy' part, electricity and today's leading technologies, wind and solar. Today they represent less than 2 per cent of global primary energy combined. Wind and solar are intermittent sources of generation, available only when the wind is blowing and the sun is shining, and therefore cannot provide energy whenever it is needed. They therefore need to operate in conjunction with a combination of so called 'dispatchable' solutions and resources that the power grid can rely on to quickly match supply and changing demand. Massive storage, imports, if and where available, of renewables such as hydro and pairing with other dispatchable sources of generation such as gas and nuclear will be required, but some system operators have estimated that the reserve margin, the amount of extra resources needed to ensure system reliability at times of stress, would need to increase from some 15 per cent to up to 300 per cent in a renewable scenario.[15] (A *Dunkelflaute*, literally meaning 'dark doldrums' or in meteorological circles, known as an anticyclonic gloom, is a term used in the renewable energy sector to describe an extended period of time, longer than a day, when little or no energy can be generated by wind or solar.) Without careful planning and an emphasis on efficiency, by shifting to wind and solar we could, in practice, be adding to the scale of the problem in the short term before we reduce it.

The second problem is that by building energy policy too largely on the premise that fossil fuels will be displaced, we simply ignore the implications of what happens if, which seems overwhelmingly likely, they are not. The risk is that we build a new energy policy that is simply not fit for purpose as it contemplates an energy system that is completely different from reality at any given time. Without applying a time basis and phasing the energy transition so that we are 'renewables ready' and we don't waste most of it, we pick the wrong solutions.

The Role of Gas

'If you have built castles in the air, your work need not be lost; that is where they should be. Now put the foundations under them.'

Henry David Thoreau

Building climate and energy policy on the premise that fossil fuels will, at the same time, be rapidly displaced is fundamentally flawed. Conventional or fossil fuels dominate and underpin the world's energy system. They currently provide the only way that power and heat get where they are needed in all their forms. The world simply does not function without them.

Utility-scale, grid-connected renewable power technologies currently provide a limited amount of electricity for the energy system. Although economies and energy users require as much heat as electricity, this is not produced directly by the largest-scale renewable power technologies such as wind and solar. Renewable heat technologies such as geothermal and biomass are further behind in scale, can be expensive and depend on the availability of local resources as heat is harder to transport. Heat pumps offer a compelling answer because they can create 3 kW of heat for each kW of electricity that they use. But they depend on electricity, demand for which is increasing and renewable supply of which is limited. As discussed above, it is also intermittent. Pairing wind and solar (currently around 3 per cent of European energy supply) together can provide a portfolio effect, smoothing production, but reliable, round-the-clock electricity supply would still require storage. In 2021, there was only enough battery energy storage in Europe to run for the equivalent of one and a half minutes. If the scale of battery energy storage increases 10 times by 2030, there will be enough storage for only 12 minutes. The time, cost and demand on resources to scale up to the European Union's target of 45 per cent renewables by 2030 are enormous and beyond levels considered technically and economically feasible. However, even if it was achieved, renewable power would still need to work alongside conventional generation. And meanwhile, demand is increasing. The demand for energy is expected to increase by 50 per cent by 2050, meaning that focusing on the supply side alone leads to only one conclusion – more of everything, including carbon emissions, and a depletion of the carbon budget compared with 2020.[16]

The need to scale up, and the scale of government support to do so, creates eye-watering growth statistics and a major business opportunity for suppliers. In Europe alone, the EU's REPowerEU initiative targets a ramp-up to 45 per cent renewables, which drives demand for battery energy storage by 2030. This is compelling for generators and investors. The EU has pledged investment of some €300 billion to support its goals. Stimulating the supply side creates new products and services, allowing generators and their supply chains to sell more products and services. This offers the prospect of substantial profits for business and job creation and other economic benefits for countries. But it will also drive demand for more fossil fuels.

One of the most important markets for renewable energy in the world has been Germany, driven by the *Energiewende* policy, now in place for the last decade. By 2020, clean energy sources such as wind, solar, hydroelectric and biogas plants reached a 46 per cent share of power generation, nearly equalling that of coal, gas, oil and nuclear combined. But energy transition is proving stubbornly difficult. In 2020, Germany's annual carbon footprint of 9 tonnes per person was around 50 per cent higher than that of France, Italy or Spain. Three years later in 2023, Germany's Chancellor spoke at Brandenburg of 'a time of great upheaval' because of the Russia–Ukraine crisis and because of the transformation required by the environmental crisis. He promised at Davos to go at 'new German speed' to reach climate goals. New laws prioritized expanding wind (planning to build four or five new wind turbines every day until 2030) and solar, at the same time as electricity and hydrogen networks. The new plan envisaged 80 per cent of electricity production coming from renewable energy by 2030.

However, the move to a power system largely dependent on weather was hitting new limits and was posing new challenges. Years of surplus generation tested these limits as much as years of shortage. Stormy days with high wind threatened to overwhelm the power grid and protracted periods of excess power in the grid from high-wind generators pushed prices negative, forcing grid operators to pay customers to take electricity on the one hand, and disappointing generators on the other hand. A low-wind year in 2021, combined with low levels of conventional energy storage, also tested these limits, pushing prices higher and threatening shortages. While Germany has, to date, avoided the scourge of energy blackouts (an average of 12 minutes of outages, the lowest in Europe and compared to an average of 4.7 hours in the USA in 2019), this has been due to the fact that the country has mostly just added clean energy capacity to supply over the past two decades, maintaining much of its conventional fossil fleet and some nuclear. Just as nuclear power has been imported from France, avoiding much of the political debate, so conventional energy is exported, a boon for the German coal industry. Germany remains highly dependent on natural gas, which forces a crisis with Russia's invasion of Ukraine. Arguably, 25 years of economic growth, and low inflation, had been fuelled by cheap Russian gas. Germany realized that it was unable to operate without it. Solutions involve replacing natural gas supplies first, although it will spur inflation, and replacing fossil fuels later. The environmental campaigning group Friends of the Earth acknowledges that 'gas will be like a fire brigade … there for when we need it'.[17] As Germany leant back on coal generation to balance supply needs, so the German Green Party proposed the addition of 25 GW of natural gas generation to balance the grid. The EU classification system for green energy, the 'taxonomy', delayed ruling on natural gas in 2021. The conclusion reached in July 2022 was that it would be green, along with nuclear. The EU's long-term objective to use gas as a transition fuel to displace coal was challenged by climate policy seeking to replace it with renewables, but for now, renewables can't work without it.

So, What Should We Do?

Given that climate change is a scientifically demonstrable phenomenon, a potentially existential threat and that the clock is ticking, it stands to reason that we must get our response right, and fast. Renewable energy will be an important part of this and offers a massive commercial opportunity. Over the next three decades, the total required investment in the energy sector has been estimated to be US$3.5 trillion per year. Electricity's share of total final energy demand could rise from 20 per cent today to over 60 per cent by 2060, implying an almost fivefold increase by mid-century, powered by renewables. So, if this is the battle plan, then we must understand the limits of our resources and supply lines if we are to succeed.

The energy transition is an ongoing process of replacing fossil fuels with renewable and efficient energy use. However, it must fit together with the significant structural changes needed in the energy system, both on the supply and demand side. Renewable power is a means to an end, but not an end in itself. If we are solving for decarbonization to mitigate climate change, then it needs to go hand in hand with efficiency. National and corporate targets for renewable generation must be accompanied by efficiency targets. National emissions reductions targets must also consider international supply chains. In the UK, net zero targets relate mainly to territorial emissions and, other than for aviation and shipping, do not fully account for international supply chains. Unless they do, like in Germany, good intentions can result in the export of bad action. We will constantly spill over the edge.

While lower-carbon energy must displace higher carbon energy over time, in the meantime it is at least as important to focus on the demand side of the equation, reducing the size of the problem that we are trying to solve. By making the energy system in any given location more efficient at all stages of conversion, generation, transmission and end use, we make our energy system renewables ready and our economy capable of generating more output per unit of energy input, less energy intense with a lower cost of converting energy into GDP, and more energy efficient from a greenhouse gas emission reduction perspective.

The focus on sources of energy must be balanced with a focus on uses. 70 per cent of energy is used in buildings, industry and transport. In the process of creating clean power for the grid, we must consider that the grid is not the customer but the conduit. We must not focus on the supply side at the expense of the demand side, on addition at the expense of reduction, on invested capital at the expense of returns.

Finally, at the United Nations COP26 in Glasgow in December 2021, energy efficiency scored a seat at the top table, for the first time, in the final text of the Glasgow Climate Pact, which called for energy efficiency alongside clean power generation to achieve decarbonization.

- Focusing on energy to mitigate climate change makes sense as some 80 per cent of human-made carbon emissions come from energy.
- For all the focus on renewable power to date, emissions keep rising and both renewables and electricity represent a relatively small proportion of global energy supply.
- Demand for natural gas is set to rise by 50 per cent by 2050.
- Time is of the essence; we only have 8–10 years of carbon budget left.
- The problem cannot be solved on the supply side alone.
- The scale and pace of change needed demands efficiency.

CHAPTER 4

Environmental Economics and Sustainable Growth

• • •

- Eroding capital, whether financial or natural, is unsustainable.
- How the challenges faced by the East held lessons for the West.
- China and Japan have been thinking long term for a long time.
- Are we trading the future of this generation for the next?
- Solving the carbon and productivity problems.

'If it's not commercial, it's not sustainable.' I first heard these words in 2006 in a restaurant in Hong Kong from a leading commercial lawyer. To him, they had nothing to do with the environment or climate change. To me, they provoked an astonishing revelation and had everything to do with the environment and climate change.

In 2006, I worked for a global bank that wanted to stand at the forefront, then, of action on climate and the environment. The bank, HSBC, promoted honourable action and strong values and gave US$100 million per annum as philanthropy to four leading environmental charities. It had a sustainable development department that reported to the chairman's office. But this was a one-way flow, out, and was not accompanied by the same level of investment, delivering investment returns, in. What if the supply of capital ran out? Unless there was a commercial return, would this level of contribution continue? Was it sustainable?

Investment returns are a reliable motivation. They are essential to the operation of capital markets. Capital contributions without investment returns diminish, deplete, financial resources over time. Losing money is an unpopular, finite and fated strategy that risks destruction of value and, in the financial and corporate world, incrimination, earning the wrath of shareholders and stakeholders in practice and, indeed, in law. Company directors would be sued if their actions knowingly destroyed shareholder value.

A similar concept of sustainable return is also essential to the way we manage our resources. There is a limit to the natural resources available to us. The concept of 'natural capital' was first used in Ernst Friedrich Schumacher's 1973 book, *Small Is Beautiful*, and developed by founders of the science of Ecological Economics. Natural capital refers to the world's stock of natural resources, including the land, air, water and all living things. Natural capital seeks to value the goods and services that these assets provide people, sometimes referred to as ecosystem services. These are infrastructure services without which human life and society cannot function and natural capital seeks to value them. So, what if we deplete them? In the natural environment, these assets generate returns and sustain themselves. Forests and rivers generate returns of trees and fish. Over-use can deplete stocks. A healthy, functioning environment can create a continuous supply of essential services. The alternative is unsustainable. Eroding or depleting natural capital beyond its limits, pushing the security of supply and ability to regenerate over the edge, can also be a finite and fated strategy that risks destruction of value and incrimination. This is now being increasingly tested in law. For example, in May 2021, a Dutch court ordered Shell to reduce its global carbon emissions by 45 per cent by the end of 2030, compared with 2019 levels. In March 2022, Shell's board of directors was sued for 'failing to properly prepare' for the energy transition.[1]

Achieving economic growth by eating into natural capital is the equivalent, in finance, of paying dividends from capital. Eventually the capital runs down and you can't pay dividends. You are not creating a return. You are destroying value and inviting recrimination.

Looking Around

I had spent the weeks before that important lunch in Hong Kong in Mainland China, working on building a real estate investment business for the bank. First impressions in Beijing were that I couldn't see down the street, because of the hue of pink dust blown in from the de-forested Mongolian desert. In Shanghai, the white dust in the sky and in my throat was from construction. In other cities, it was pollution from industry and coal-fired power stations. Energy supplies were stretched, and depleted aquifers left trees in the avenues leading up to so-called 'eco-towns' to die of thirst while the ground conditions could not reliably support the buildings that were meant to stand on them. (Today, the High Plains aquifer, east of the Rocky Mountains in the USA, is suffering severe depletion from irrigation of the USA's 'breadbasket', with some predicting its demise in a generation.) This seemed unsustainable. There must be a limit, but also, given the scale of the problem, an opportunity to invest in essential infrastructure services, such as cleaner and more efficient energy and water solutions. These 'environmental infrastructure' services seemed no less essential than the hospitals, schools, and roads that the traditional infrastructure investment was designed to build and operate. If environmental infrastructure solutions addressed a need and solved problems for government and society, structured correctly they could generate sustainable returns.

My efforts to launch the real estate investment business for the bank in China were successful. We sought to raise US$500 million and raised US$708 million from institutional investors from all over the world in record time before we stopped trying. I had the privileged experience of building a joint venture with a leading Hong Kong family, hiring a team, opening an office in Hong Kong and starting a large business within a large business. However, my first efforts to convince colleagues and superiors in London that the future was environmental infrastructure were not so successful. As far as I was concerned, I had seen the future. In trying to achieve economic growth so fast, bringing relative prosperity to hundreds of millions of people emerging from poverty, China was creating severe environmental degradation that was, at the same time, destroying its economy and depleting its capital base. This is what the future could look like at home.

But this was a global problem. To solve it, new solutions would need to be found to new problems. It would require complete focus, long-term thinking, determination, resilience, and some equilibrium of patience and impatience. We should, I said, set up a dedicated division of the bank to focus exclusively on environmental infrastructure, locking ourselves in a room until we came up with solutions to the problems that would define our next generation. It was a great idea, said the bank bosses, but they were not ready yet. They had just printed their millionth credit card in China and anyway, they said, there was a risk that renewable energy, which defined the topic then as much as now, was a fad that would disappear when the oil price fell below US$50 a barrel. So, I said, I would do it. On 10 October 2007, I drove from Canary Wharf to Mayfair and started Sustainable Development Capital LLP (or SDCL).

China

In 2006, China's 11th Five-Year Plan illustrated the degree to which it had been rocked by the realities of the impact of environmental degradation on its economy and society.

In a landmark paper in 2006 for the Council on Foreign Relations, 'China's Environmental Challenge', Elizabeth Economy produced a stark summary of the direction of travel that China was set on. In 1997, the World Bank had published its report, 'Clear water, blue skies', that suggested that the cost of environmental pollution and degradation in China was equivalent to 8–12 per cent of GDP annually.[2] This estimate included the costs associated with lost days of work from pollution-related illnesses, contaminated crops and fisheries, and industry closures due to lack of water, among other factors. By 2003, Chinese media were reporting that floods and droughts had led to economic losses totalling 200 billion yuan (US$24 billion); desertification resulted in direct economic losses of 54 billion yuan (US$6 billion). Acid rain cost China US$13 billion and water scarcity cost China US$28 billion in lost industrial output. China's fisheries lost US$130 million in 2004 due to water pollution, representing an increase of over US$40 million from the previous year. In terms of biodiversity, China's opening up to the international community brought with it the introduction of more than 250 alien species that caused environmental damage with a price tag of US$14.44 billion in 2000. In 2006, 25 per cent of China was desert. The World Bank had stated that annually 300,000 Chinese were dying prematurely from respiratory disease related to air pollution.[3] Tens of thousands of people participated in environmental protests, often spilling over into civil unrest.

China responded. The 11th and subsequent Five-Year Plans prioritized environmental protection and 'harmonious society'. Massive reforestation, movement of polluting industry out of urban areas, 'eco-cities', and investment in cleaner and renewable energy were all promoted and achieved at a globally unprecedented scale. However, one of the most extraordinary policy measures was a focus on energy efficiency. China demanded a 20 per cent reduction in energy demand per unit of GDP output over the 11th Five-Year Plan from 2006 to 2011, or 4 per cent per annum. And it would achieve it by hook or by crook. China understood that de-coupling economic growth from energy consumption and pollution was critical to sustainable growth. Indeed, China's National Air Quality Plan was introduced in 2013 and presaged a decline in particulate pollution of nearly 40 per cent by 2020. Air pollution in Beijing fell 55 per cent, adding an estimated four and a half years to the average inhabitant's life expectancy. China knew the limits of its largely imported primary energy supply and the effects of environmental degradation and it was already on the edge.

China was starting from a low base in 2006. Low use of energy efficient technologies in China's buildings and industries placed heavy reliance on use of coal and generated pollution. China's buildings consumed two to three times the energy used by developed countries in comparable climates, and depending on the industry, Chinese factories were three to ten times less efficient than their Indian, US and Japanese counterparts.

While this gap has probably narrowed to two to four times and China now generates around 17 per cent of its energy from low carbon sources, China still relies on coal for some 55 per cent of its energy consumption.

President Xi Jinping announced in 2020 that China would be carbon neutral by 2060. Major progress on decarbonization is being made, including a massive expansion of wind and solar. In 2022, China's new renewable energy capacity additions were approximately equivalent to all those in the USA, Europe, India, Southeast Asia and Latin America combined. Solar panel installations are growing at a pace that would increase global capacity by 85 per cent by 2025. A 2023 report by Global Energy Monitor entitled 'A race to the top' predicted that China's 2030 green energy targets were on track to be exceeded 5 years ahead of schedule. It pointed out that, as it went to print, China had installed more solar panels in large-scale projects than the rest of the world combined.

However, the renewable surge and curtailment of coal-fired power generation found its edges in the summer of 2021 when energy shortages accompanied increased demand from the first phase of economic recovery from the COVID-19 pandemic. The energy crunch convinced policymakers of the dangers of moving too quickly away from coal. In the first six weeks of 2022, China approved 7.3 GW of new coal-fired power plants, double the figure for the whole of 2021. In 2022, China built approximately two new coal-fired power stations every week, often co-located with wind and solar generation for backup and balancing. By the end of 2022, China had approved 106 GW of new coal capacity, equivalent to approximately all the coal capacity that the USA and Europe had decommissioned in the previous five years.

While coal still represents 60 per cent of China's electrical generation, only 6 per cent comes from natural gas. However, China's need to reverse this ratio lurks behind Russia's 'Pivot to the East', to satisfy this demand for natural gas to replace coal.

Indeed, in conjunction with energy efficiency and appropriate renewable energy generation, one of the most impactful changes that China could make would be to displace coal with gas. This shift was responsible for the USA being able to claim the largest reduction in CO_2 emissions of any nation in the 2010s. China is the largest consumer of coal in the world. Each year since 2011, China has burned more than half of all the coal used worldwide, compared, for example, to the next largest users: 12 per cent for India and 8 per cent for the USA. Switching from coal to gas would deliver massive greenhouse gas emission and pollution cuts.

Japan

Japan's energy policy is based on the principles of three 'Es' and an 'S': energy security, economic efficiency, environment, and safety. An interesting and practical alternative to the acronym 'ESG': environmental, social and governance, the little-understood acronym dominating the corporate and financial market narrative in the West.

At the same time, the limits of Japan's natural resources had pushed it, for the lack of any other choice, in the direction of resource efficiency since the oil shock of the 1970s. From 1973, Japan effectively de-coupled economic growth from energy consumption, through a series of supply- and demand-side measures, including energy management and efficiency. GDP grew over 2.6 times between 1973 and 2019, with a corresponding increase in energy consumption of only 1.2 times. Energy used per unit of GDP output roughly halved over the period. Japan's primary energy consumption, divided by its GDP in 2021, was 1.0, compared to 1.3 for the EU, 1.5 for the OECD, 1.8 for the USA and 4.3 for China.

Japan's efficiency measures have directly contributed to an actual reduction in greenhouse gas emissions. An example of an effective measure on the demand side was a revision to the 1998 Energy Conservation Law, which introduced a 'Top Runner' programme to incentivize efficiency in transport and home appliances such as air conditioners and refrigerators. A label was awarded to products that exceeded the performance of the previously most efficient product made, allowing consumers to make smarter purchasing decisions. The programme delivered a 6 per cent reduction in greenhouse gas emissions between 2013 and 2019. In industry, co-generation, recovery of wasted thermal energy, high-efficiency furnaces, high-efficiency boilers and facilities to use steam effectively all made a difference – industrial emissions in Japan have actually fallen by 20 per cent since 1973, not increased at all. In the residential sector, Japan has developed highly efficient thermal transfer technology, otherwise known as a 'heat pump', widely applied to air conditioners, refrigerators, water heaters and other devices. CO_2 emissions in the residential and non-domestic sectors dropped by 21 per cent between 2013 and 2019. Under the new Strategic Energy Plan, a target has been set to reduce CO_2 emissions from buildings (by 56 per cent), industry (by 38 per cent) and transport (by 35 per cent) by 45 per cent by 2030, which will require a 23 per cent reduction in final energy consumption. In 2022, the Building Energy Efficiency Act was revised to extend mandated energy standards to all residential and small non-residential buildings and to extend the Top Runner programme to large house builders.

Japan is the world's fifth biggest carbon emitter, but it has set ambitious targets for decarbonization, seeking to achieve net zero by 2050 and a 46 per cent reduction in carbon by 2030. This is flatly against the odds, but characteristic for a leader in energy efficiency. Japan runs 90 per cent on fossil fuels, with limitations on its renewable energy capacity from wind and solar given its small size, high density, topography and deep waters. Japan is the largest importer of renewable natural gas in the world; 90 per cent of its energy comes from oil, gas and coal, substantially all of which is imported from the Middle East, Southeast Asia and Australia. Yet it has set a course to lead on hydrogen, setting up a trade corridor with Australia, ammonia, new nuclear (particularly small modular reactors) with safety measures, coal and diesel to gas conversions and to promote public transport. Prime Minister Fumio Kishida has said that state and corporate spending on green technologies, including the decarbonization of industry, will reach US$1.1 trillion over the next decade. The *Financial Times*, Nikkei Asia and Statista compiled an 'Asia-Pacific Climate Leaders' list, comprising businesses that have significantly cut their 'Scope 1' (from own operations) and 'Scope 2' (from energy purchased) greenhouse gas emissions relative to revenue.[4] Out of 200, 86 were from Japan, far more than any other country.

Japan recognizes that it has no silver bullets and, given its current reliance on fossil fuels and its resource constraints, it has some good bullets but none of its bullets is perfect. Its energy policy has been largely paralyzed since the Fukushima nuclear disaster in 2011, following which it shut down most of the nuclear reactors that had supplied around a third of the country's electricity. This deepened its reliance on fossil fuels. But while its resource constraint is more severe than that of Europe in some ways, in others it is similar: 90 per cent versus 80 per cent fossil fuels. The perceived as well as real constraint, which has fewer solutions, has however forced it to innovate.

The importance of energy security and energy efficiency was brought home in June 2022 when the Japanese government was forced to call on businesses and the public in Tokyo to cut electricity use. Lack of generating capacity risked a power blackout for the second time in the year. The power crunch was exacerbated by war and the climate. Curbs on Russian gas imports were caused by the war in Ukraine. Temperatures in the Tokyo area rose above 35°C following a record early end to the rainy season, causing a surge in demand for electricity for air conditioning. Good energy policy is key to keeping the lights on and the economy running.

The Big Picture

National examples help to illustrate the real, day-to-day interaction between resource efficiency and economic performance and productivity. It is relatively easy to identify and measure what has worked and what hasn't.

By contrast, the more broadly we zoom out, the more room for error and debate. Economists heavily caveat every major study on the impact of climate change on GDP. Climate economics is, they say, inherently uncertain. There are an extraordinary number of variables, interactions, feedback loops and unknowns. To that, you must add politically and philosophically coloured filters to assign value, for instance based on the discount rate that you apply to a problem that may occur with a given probability in the future. And when you make projections about value (including returns), you have to take into account the allocation of capital investment, towards prosperity-based outcomes, which some argue can be in conflict with climate ones. According to the Sixth Assessment Report of the UN IPCC, leaving aside the economic benefits of mitigating the costs of damages from, or adapting to, climate change, GDP is projected to be slightly lower in 2050 under the current climate change policy framework. However, the global economic gains associated with being able to limit warming to 2°C are projected to exceed the associated mitigation costs. Mitigation is expected to more than pay for itself.[5]

There is also an inherently austere side to climate economics. If we solve only for GDP, then we tolerate impacts on life that don't show up in the numbers and fail to value losses associated with them. There is a real risk, according to the IPCC, that a billion people will be living in extreme, potentially fatal, heat conditions by the 2030s and that hundreds of millions will be displaced in the meantime due to environmental factors, noting that the 2022 number from the United Nations for the displaced has already reached 100 million. It can be argued we can sustain losses of 5 per cent of GDP in present or future value from climate impact, but which of us are we referring to under these scenarios? I would not dare to talk about GDP to a widow who had just lost her family to a resource-driven conflict or to a climate-induced natural disaster. GDP also does not count plants and animals, the land and oceans and biodiversity, on which societies depend for survival, where if these resources become over-exploited then they deplete and societies fail, falling over the edge.

Searching for Solutions

Economists have articulated alternatives for sustainable economics, often as a reaction to 'unsustainable' consumption-based GDP growth and, on the other hand, recognition of negative feedback loops, such as reducing demand for fossil fuels by increasing the share of renewables, in turn reducing the price of fossil fuels and increasing demand for fossil fuels. How to turn a vicious cycle into a virtuous one? How to deal with the 'tragedy of the commons', which results from individuals acting in their own interests, but which are not in the interests of others because they deplete or ruin shared resources? The consequences of these actions that affect people who did not cause or benefit from the actions are often referred to as 'externalities'.

The first and most common solution proposed is to price the externality, or to apply what some refer to as the 'polluter pays' principle. Polluters can pay via a carbon tax or a levy or a traded instrument, such as a compliance certificate or credit, but either way, success depends on a globally coordinated and consistent system to derive an effective price for carbon. However, even the strongest proponents of a carbon tax have labelled the chances of global coordination in line with an optimal climate policy a 'fairy tale', given that it would require politicians to enforce, however rational, increasingly expensive taxes covering everyone on the planet over the next 80 years, during which time the USA, for example, will have had 20 different presidents, not all of whom may have exactly the same point of view about climate change and how to respond to it. So far, there is no more consistency on the more market-based approach to compliance or voluntary certificates, where you pay to pollute, sometimes buying the right to do so from another party. These may be effective but can also be controversial in terms of standards and additionality (i.e. whether projects they fund do any good versus business as usual) and, at the time of writing, certificates in one part of the world often cost ten times those in another part of the world, for the same tonne of carbon.

Carbon

It's interesting that pricing carbon is almost always the first solution proposed and has been ever since I have been actively involved in climate finance, from 2006. I think that it's just as interesting that no one agrees on what the second and third solutions are. A contender for the number two spot is public sector investment, which can come in many forms. Public sector investment is often directed to address public goods, from which a societal rather than a commercial or financial return can be derived. Public ownership of infrastructure networks such as rail, road, electricity, water, waste, and communications systems is often cited, partly because they have universal service obligations, linked to the public need to have excess capacity to ensure security of supply. R&D has also been held out as a public good that requires public sector investment. However, there is also a case that public infrastructure can be delivered most efficiently in partnership with the private sector, provided that the private sector is sufficiently incentivized to deliver on time and on budget and to maintain operations to required levels of service. Private sector capital abhors a loss. If public sector makes infrastructure investment for private sector commercially viable by bearing the (mostly demand) risks that it cannot, and by providing long-term certainty of revenue, then private sector capital at extraordinary scale can and has been unlocked to pay for the infrastructure and spread the cost of its useful life. Public sector gets high-quality infrastructure on time and on budget and with risk transferred to contractors. Private sector gets a reasonable risk-adjusted return for the capital and expertise it deploys. This is a sustainable and scalable model, with few limits.

Productivity

Of course, if public and/or private sector capital can be invested in projects that cut costs through productivity improvements, like energy efficiency investments for instance, then the realized benefits, which are measurable and verifiable, can translate into more free cash flow for the party now paying less for its infrastructure service, on the one hand, or on the other hand, returns that can be reinvested by the party investing the capital and earning its returns from, essentially, savings. This can also substantiate the 'additionality' or 'net environmental gain' that economists seek to demonstrate to ensure that consequences of investment deliver demonstrably more 'public good' than otherwise would have happened anyway. The extent to which a public–private partnership as described above, or any collaboration between public and private sector described in this paragraph, is feasible depends on the relationship between public and private sector, which in many countries, including the UK, is often one of suspicion of the private sector on behalf of the public sector. This limits the penetration of goods, services and investment into the public sector, via procurement roadblocks, accounting complexities and a preference for cross-subsidies funded by public borrowing.

Future Generations

There are many other debates in environmental economics that are, in essence, partisan in that support tends to depend on either side of the political spectrum. Some economists argue that the current generation must face up to its actions to protect future generations and pay now for the benefit of future generations (tax and spend) or borrow and spend (which in fact puts cost on future generations), in each case involving the government acting to correct a market failure, both on the financing (sources) and delivery (uses) side. Some proposals involve assignment of property rights over public goods or shared resources so they can be managed (like the enclosure movement) but this has obvious challenges on a larger or global scale and defies the propensity of nation states to compete for resources actively, and sometimes aggressively.

Governments and companies, through international cooperation platforms and diplomacy, including the annual climate conference hosted by the United Nations, are aiming at the objective of shared management and regulation to ensure that scarce resources are exploited prudently, with some encouraging signs of success. There were low expectations for the Glasgow COP26 in 2021, but one of its more remarkable achievements was the series of coordinated announcements for countries to stem deforestation, reduce methane, phase down coal and make cars net zero emission. The USA's 'Inflation Reduction Agreement' proposed to levy a fee on the methane emissions of oil and gas producers and pipeline operators. Methane has up to 80 times the warming potential of carbon dioxide over a 20-year period and up to a third of the 1.1°C warming since pre-industrial times has been attributed to it. The USA is the second largest source of methane emissions after China and ahead of Russia.

Over 200 countries signed up to the ambition of limiting global temperature rise to 1.5°C, and over 1,000 cities to halve emissions by 2030. Nonetheless, harmony had been broken within months. European countries lurched back to gas and coal in the face of an energy security crisis following Russia's invasion of Ukraine and commitments to net zero emission cars were rejected. UN climate talks in Bonn in June ended in acrimony, with poorer nations accusing the USA and Europe of betrayal over the issue of finance to combat climate change. By October, the International Monetary Fund (IMF) was moved to interject, pointing out that any short-term economic costs would be 'dwarfed by the innumerable long-term benefits of slowing climate change'. That the UN referred to a 'wasted year' since COP26 as Egypt was hosting COP27 speaks for itself.[6]

Nevertheless, we have an enormous advantage over past generations. This is not just a dramatically improved benefit of hindsight but also one of foresight. Based on our ability to measure and assess, we can predict, with some reliability, scenarios for the future. The first chapter of this book looked at history seemingly repeating itself, but also explained this illusion in the context of causes and effects. We can apply similar logic to the climate problem and prioritize solutions based on the effect or impact that they should make.

Before we look at solutions, I repeat the statement: if it's not commercial, it's not sustainable. This doesn't mean that things that are not commercial cannot be done, it just means that they cannot be done reliably or at all by the private sector and may depend for financing and implementation on a party that values public good, so the public or philanthropic sectors. So, the big question is where to start and what to focus on, what is the plan? The first responder in the climate crisis is usually identified as renewable energy. We will cover renewables in some detail in the next chapter, but we will also keep in mind another David MacKay paraphrase, 'every big thing helps'.[7] We will consider, in context, other solutions for public and private sector, such as efficiency in the use of energy and other resources, decentralized energy and the role of digitization, governments and companies. We will try to benchmark ourselves with reference to economic productivity, efficiency and commercial sustainability, a sustainable growth paradigm. We will look for solutions that make society better off rather than depending on sacrifice, whether shivering at the thermostat or shuddering on the thresholds of conflict.

We will keep in mind the fact that, at the end of the day, we are doing all of this for people, to make a better society, so we will, if we can, avoid placing the blame at the feet of individuals, which can be a tendency of liberal democracies. We will recall that in both the East and the West, which can have radically different political and social philosophies, the rules and actions are largely set and taken by institutions, and this is where the biggest differences can often be made.

However, radically different political and social philosophies can also result in different solutions proposed for the same problem.

How

In this and many other ways, the debate is moving on from 'what' needs to be done to 'how' and in what order of priority. Most share of voice, effort and money goes to generation of energy, while there are legitimate questions as to how we use energy, demand versus supply, reduction versus addition, productivity versus consumption. In parallel, the questions are about how carbon should be removed from the atmosphere (for instance through reforestation and restoration of natural capital, together with carbon capture, utilization and storage) and how this can be funded or even commercialized.

We might also recall the principle of the 'golden mean', originally proposed by Aristotle, Confucius, and other classical philosophers, as a warning against leaning on extremes. The Pareto principle or 80–20 rule explains this in numbers. For example, it can take 20 per cent of the time to complete 80 per cent of a task, while to complete the last 20 per cent of a task can take 80 per cent of the effort. We need to recognize if we are spending 80 per cent of our effort on 20 per cent of the task and adapt accordingly. Achieving absolute perfection may be impossible and so, as increasing effort results in diminishing returns, the same activity becomes increasingly inefficient.

- China was losing 8–12 per cent of GDP per annum from environmental degradation in 2006.
- China chose to reduce energy consumption per unit of GDP output by 4 per cent per annum.
- Today, China needs to substitute coal for lower-carbon solutions, namely natural gas (on which it will rely on Russia) and renewables, and to focus on efficiency.
- Japan has been a leader in efficiency.
- Sustainable development = resource efficiency.

CHAPTER 5

The Renewable Energy Paradox

• • •

'Politics is the art of the possible, the attainable — the art of the next best.'

Otto von Bismarck

- Focus on the best available technologies in the time available to implement them.
- There is no zero carbon or limitless energy source; every energy generation technology has its limits.
- We need to balance and diversify sources of energy supply.
- And we can't just add to the system – we need to reduce.

Sir Robert Alexander Watson-Watt, a radio technology pioneer who developed the early warning radar for the Royal Air Force to counter the rapid growth of the Luftwaffe, a system that helped win the Battle of Britain, promoted a 'cult of the imperfect', summarized as: 'Give them the third best to go on with; the second best comes too late, the best never comes.'

While politics is sometimes about second best, so with power. As Voltaire said, 'the best is the enemy of the good'.[1] Put another way, planning carefully and realistically, we must avoid sacrificing the good on the altar of the perfect. So-called 'transition' fuels, such as natural and recycled gas displacing coal, do have a role to play, particularly if they can be designed to accommodate lower-carbon fuels when they are available at scale. Some more 'conventional' solutions may also be needed to accommodate and balance renewables in the short to medium term, while the system changes are implemented. In the meantime, the process of decentralizing and focusing on energy efficiency may, in practice, have the biggest impact in the short term while we transform our energy system to renewable energy in the longer term. Otherwise, we risk sacrificing major, achievable, and enduring achievements and efficiencies now in favour of planned, hoped for, or even imagined improvements in the future. We will not win if our plan depends on solutions that will simply come too late for the problem.

There are technical, economic, and physical limits to the generation of renewable energy and the ability of the energy system to absorb and manage it. Time is another key factor. While it is highly probable that many of these limits expand significantly over time, the timeframes involved to deploy renewable energy at a global scale that substantially displaces fossil fuels will take several years at best and, most probably, decades. Within the context of the potential depletion of our carbon budget by the end of this decade, urgent energy security problems affecting the lives of hundreds of millions and energy costs that matter for billions of people, time may in fact be the key. While we must plan and invest for the future now, we also have to accept that renewable energy will help less with today's problems and more with tomorrow's. We must take time into consideration and, in the meantime, use the best available, and most efficient, technology today on our journey through the energy transition.

Solar – Light and Its Limits

In 1905, Albert Einstein wrote a paper, *Concerning the Production and Transformation of Light*, that would ultimately earn him the 1922 Nobel Prize in Physics for 'his discovery of the law of the photoelectric effect'.[2] The Greeks and Romans had amplified and reflected the sun's rays to create fire, but the solar photovoltaic effect was defined by 19-year-old Edmond Becquerel in his father's lab in 1839. However,

Einstein's predecessors in solar photovoltaic technology development, such as Charles Fritts and Aleksandr Stoletov, had limited success in achieving much more than 1 per cent efficiency. Einstein's work laid the theoretical foundations for the first practical solar cell developed in 1954 at Bell Laboratories by Daryl Chaplin, Gerald Pearson and Calvin Souther Fuller. By 1958, a solar array appeared on the *Vanguard 1* space mission and solar went on to become the default source of energy for space application.

Solar energy, in theory, can provide more than 10,000 times the energy needed by the entire planet. While it is one of the largest, cleanest, safest and most cost-effective sources of renewable energy available, it has its limitations. Most solar cells available on the market today operate with efficiencies of between 15 per cent and 23 per cent. Traditional single-junction cells have a maximum theoretical efficiency of 33.16 per cent, the Shockley–Queisser limit. In 2019, the world record for solar cell efficiency at 47.1 per cent was achieved by using multi-junction concentrator solar cells, developed at National Renewable Energy Laboratory, Golden, Colorado, USA. The amount of electricity that can be generated by a solar panel is usually calculated as a function of efficiency and capacity, which depends on the proportion of the day that is sunny. The capacity factor is the ratio of energy generated over a period divided by the installed (nameplate) capacity. This is generally 25 per cent for solar and wind, compared to 40 per cent for hydro and combined cycle gas, 70 per cent for coal, 85 per cent for CCHP and 90 per cent for nuclear.

However, in practice, as I am finding in developing and investing in both grid-connected and rooftop solar, there are several other key limiting factors in land, time, planning, supply chain, grid integration and access to the resources needed for construction. Starting with land, the space that would be required to power the USA with solar would be around 1 per cent of the territory, 15 million acres, all the space occupied by roofs and roads. (Europe is making solar on all new roofs mandatory.) Six times the space would be required to power the country with wind. Time to develop projects, including securing real estate rights and grid access, can typically be one to three years, with a construction period of nine months to a year. To construct solar panels takes energy, commodities, copper and minerals sourced from a small number of countries, many of which are not known for the highest environmental and social standards. Panels need water to clean them and can degrade over time. The intermittent nature of solar power generation can cause market dislocations with integration with the grid, which can require subsidies to correct them, and larger utility-scale projects built away from the point of use can involve substantial transmission costs, which can be higher than conventional generation because line capacity must be higher because of the lower capacity factor.

The Way the Wind Blows

We need renewables to decarbonize in the energy transition and they are an integral part of any 'net zero' carbon plan. However, they are not zero carbon, when accounting for the energy and other resources needed to build them, operate them and, in due course, decommission them. They are lower carbon, but not zero carbon. They can also have other forms of environmental impact. A quarter to a half a million birds and over 600,000 bats were killed in 2012 by wind turbines in the USA. More birds die flying into windows and cars or being caught by pets, but it is still a large number. This could potentially be mitigated by curtailing generation during migration season, but we are also learning about other environmental issues during operation, such as the combined impact on wind velocity and turbulence and therefore temperature. One study argued that if all electricity demand in the USA (0.5TW) was generated by wind power, the continental USA surface temperatures would be warmed by 0.24°C.[3] Offshore wind, where wind speeds can be higher, larger turbines can be installed and major, gigawatt-scale projects can be developed, is big enough to make a difference and is well established in Europe. It does, though, depend on resource availability, water depth, birds and mammals and time to plan and construct.

Hydro

The challenges and limits of generating more large-scale clean energy – and not just electricity but also heat – demand diversification beyond wind and solar. Hydroelectricity is one of the largest sources of renewable energy and represents around one-third of Europe's installed capacity. There are significant environmental issues associated with interference with rivers, resource and political issues associated with the construction and operation of dams and, seemingly increasingly, climate-related issues with low rainfall or droughts in Spain and California in 2021 and 2022 causing shortages and risking blackouts.

Hydrogen

Hydrogen has enormous advantages in terms of low emissions (provided, as it turns out, that it is not leaked into the atmosphere), the ability to generate industrial temperature heat, which traditional renewables cannot, and to serve important applications in decarbonizing long distance transport (some 8 per cent of global emissions) but it is rarely an efficient means of generating electricity and depends on what is fuelling it, i.e. wind and solar (with energy losses of some 60 to 80 per cent by turning sunlight into hydrogen, then ammonia and back to electricity) or more conventional (natural gas or even nuclear) power via electrolysis. Hydrogen has become one of the most exciting 'near-zero' greenhouse gas clean energy stories for politicians and the media, offering the prospects of a clean fuel that emits only water vapour and warm air. Different labels are given to hydrogen produced from different sources, such as 'grey', produced from natural gas, 'blue' produced from natural gas with carbon capture, 'green' produced from renewable energy and 'pink' produced by nuclear. Practical feasibility is yet to catch up to some plans or hopes for hydrogen. The EU's 'Fit for 55' package targets 10 million mt/year of green hydrogen production, which would require 477 TWh of renewable power generation. That is almost the entire EU solar and wind generation in 2021.

Bioenergy

Bioenergy and synthetic fuels such as ammonia and methanol have the advantage that they can be used as 'drop in' fuels in conventional combustion engines, which helps mitigate the fact that much infrastructure for the next decade is already 'locked in'. In the 'REPowerEU' package, the European Commission called for an increase in the share of hydrogen used in industry to 75 per cent by 2030, up from its previous ambition of achieving a 50 per cent share announced in 'Fit for 55'. Achieving those numbers will be difficult not only due to electrical demand, but also reinvestment cycles. For example, recent analyses of potential hydrogen demand at industrial parks in Northwest Europe, together with a timeline for logistical incorporation of the fuel, have found that only a minority of the potential hydrogen demand from industrial plants could be delivered by 2030 without replacing plants before the end of their lifetime. According to the IEA's Hydrogen Projects Database, at the end of 2022, less than 5 per cent of the planned capacity by 2030 has reached a final investment decision or is under construction. Over 95 per cent of these projects are in various early stages of development, i.e. either conceptual or subject to feasibility studies.[4]

Renewable Heat

Geothermal requires time, drilling for discovery of resource and careful management to ensure that emissions are controlled. The EU is, separately, targeting the installation of 50 million heat pumps by 2030, a gigantic task considering only 17 million heat pumps were installed across Europe in 2022. Installing that many heat pumps could have major impacts on residential and commercial emissions and energy usage – if an average house installs a heat pump instead of a gas furnace and water heater, its heating emissions decrease more than 45 per cent over ten years. Heat pumps in industry could be a key to electrification and decarbonization.[5] However, a practical concern is that they do not operate as efficiently in all temperatures and may not necessarily yet produce cheaper energy, although innovation is making heat pumps more efficient and temperature flexible. Additionally, solutions like a shared ground heat exchange, which is a distributed heat pump combined with an ambient temperature network, could improve efficiency and uptake.

Nuclear

Nuclear is often hailed as a baseload low-carbon energy source. Nuclear power stations are however fuelled by a non-renewable resource, uranium. While relatively abundant, it is procured from a small number of states with varying levels of governance. Nuclear is also intimately connected with the oceans as it needs access to cool water, which is increasingly challenging. Two-thirds of the US oceans are already warmed by power stations. Installation of new capacity can take decades and the bespoke design of large-scale projects has contributed to a poor track record in the construction phase, with most projects running years late and billions over budget. Small modular reactors offer a solution but not before the end of the decade.

Safety is another key issue. Nuclear has secured many of the most notorious headlines, given major disasters that remain fresh in living memory, such as at Chernobyl and Fukushima. However, nuclear, analogous to flying, generates some of the highest levels of anxiety but lowest number of casualties. Fewer people have in fact died from nuclear accidents than from any other major energy generation source. Nuclear waste, also a contentious issue, can be accommodated by technology as mundane and proven as ten inches of concrete, behind which you would experience less radiation than a banana, a Brazil nut, or a cigarette. Deaths are by no means the only measurement of safety, but the appalling track record of coal in creating water and air pollution, acid rain and land degradation attests to the severe damage that some forms of generation can do much more silently to people, animals and habitats.

Transport Fuels

Particulate emissions from transport fuels are a key safety issue. This is potentially one of the most compelling arguments to shift cars and other passenger vehicles such as buses to electricity from petrol and diesel. More people die prematurely each year from air pollution in cities than they do from road traffic accidents, from COVID-19, or, for that matter, from murder, terrorism and war put together. However, whether the electric vehicles are substantially lower carbon depends to a large degree on the carbon intensity of the local grid from which they are pulling power. If the grid relies on fossil fuels, for example, then the case for decarbonization becomes more difficult once the lifetime emissions of a car are considered, with 90 per cent or so of the energy and carbon involved typically being used to make or destroy the car, rather than drive it around. By contrast, aircraft emissions come, mostly, from flying.

Intermittency

While there are limits linked to resources and competing priorities to balance, another form of balance has to do with accommodating intermittent energy sources such as wind and solar into the grid. The power is only available at times that the wind is blowing and the sun is shining. This creates significant challenges for grids as they need to be able to supply energy whenever it is needed, rather than when it is available. Supply and demand determine pricing and when high levels of renewable energy supply hit the market at a time when there is no demand for them, prices can go negative (as felt acutely in Germany, resulting in the financial collapse of some utilities) and power output can be reduced or 'curtailed'. This can result in large amounts of renewable energy being wasted. New records are a frequent occurrence. For example, in the UK enough renewable power to supply 800,000 homes was wasted in 2020 and 2021.

This cost a record £507 million in 2021, contributing to higher emissions and energy bills for consumers. This is not a very British problem. The Californian market is a perennial victim, while in

China, more wind turbines and solar panels are being added than the grid can digest, such that in Inner Mongolia, for example, nearly 12 per cent of wind power was wasted, along with 10 per cent of solar power in Qinghai. The key reasons for the wastage are lack of storage, conversion, and long-distance transmission infrastructure. China's investment in these solutions will be crucial to avoid a slowdown in clean energy adoption by the world's biggest emitter. For the UK, saving this cheap waste power to be used later or using it to produce green hydrogen will be a key consideration as the government plans billions of additional investments in offshore wind capacity to 50 gigawatts by the end of the decade. The problem has another awkward political dimension. Although it is normal to pay turbines to switch off to balance the grid when it cannot use the wind energy being produced, these charges are paid for via consumer bills, which is hard, particularly in a cost-of-living crisis.

Batteries

Battery energy storage is promoted as a balancing technology, allowing energy to be stored, on the supply side, until it is needed and, at the point of consumption, to shift demand on the grid to when energy is most abundant and best value. Existing battery capacity in Europe and the USA is, however, only capable of supporting the grid for a matter of minutes, and this will remain the case if we assume it will be scaled up ten times. Issues of resource use and time apply. 'Long-duration' storage technologies include established solutions such as above-ground pumped storage hydropower, while new mechanical, thermal, electrochemical and chemical solutions are being developed. Additional flexibility can be provided by flywheels and other technology that creates a 'spinning reserve', capable of generating electricity for the grid at the right frequency and quickly, a role traditionally served by gas. The role of gas in the energy system is still as important today in renewable-heavy grids such as Ireland's for balancing as it is in other markets for front-line power generation.

The Limits of Renewables

In summary, renewable energy is not a silver bullet. We will need a diversity of energy sources and technology selection depends heavily on local resources and geography. But it is also not a silver bullet because, taken as a whole, the scale of the increase in capacity additions required for renewable energy to dominate is too large and would take too long, on its own, to achieve either the rapid decarbonization needed for a 1.5°C pathway or the improvements in energy security and resilience needed for the modern world and geopolitical landscape. A spokesperson for BlackRock, the world's largest investor, put it like this: 'A few years ago, everyone was very focussed on "let's build more renewables" [but now] … everyone's realized that … we can't get there without decarbonizing all these other hard-to-abate segments.'[6]

The limits to renewable energy growth rates challenge the premise that it alone can mount a timely response to global climate change. Adding new renewable energy capacity is needed but we must be realistic about how far and how fast they can take us. The future is renewable, but the present is not. Wind and solar represent 1.3 per cent of the world's primary energy, versus 82 per cent from fossil fuels. In the USA, renewables represent some 11 per cent of energy consumption today and are expected to reach almost 16 per cent by 2050, half of which would be wind and solar. Wind and solar supply renewable electricity to the energy system. However, electricity itself represents only around 20 per cent of the world's energy system.

The landmark Inflation Reduction Act aimed to support the reduction of carbon emissions by 40 per cent and to increase the share of clean generation in the electricity system from to 40 per cent to 60–80 per cent by 2030. While in some ways it provided something for everyone, it provided most of its support for wind and solar. And it did not disincentivize the use of fossil fuels. It still focused on the supply rather than the demand side of energy. The clean technology provisions in the Act are expected to lead to only very small reductions, maybe less than 1 per cent, in petroleum consumption and reductions of 3–10 per cent in natural gas consumption.

Meanwhile, in the Asia-Pacific, coal generation rose by 6 per cent in 2021 and it is being produced by a relatively young fleet of power plants that will be in service for many years to come. Progress to date does not constitute a displacement of fossil fuels. We need to accelerate change, but it will take time and innovation and creativity and investment, together with a realistic strategy to get there and a plan as to what to do in the meantime.

During this time, we will have to confront and manage the reality that the energy system is, and will remain, dominated by fossil fuels. Even continued cost reductions for renewables won't, per se, spell the demise of conventional fuels and a displacement of use. Cost reductions will apply to fossil fuels too. Shale and other non-conventional oil and gas are getting cheaper to produce. To the extent that renewables, which have very low marginal cost, are successful in displacing demand for oil, they will be competing with low marginal cost of oil production, US$5 or less in Saudi Arabia, versus a much higher world price.

Renewables will also be competing with traditional fossil fuels for energy security. It's a doubled-edged sword. A recent article in *McKinsey Quarterly* noted: 'The 19th century's naval wars accelerated a shift from wind to coal-powered vessels. World War I brought about a shift from coal to oil. World War II introduced nuclear energy as a major power source. In each of these cases, wartime innovations flowed directly to the civilian economy and ushered in a new era. The war in Ukraine is different in that it is not prompting the energy innovation itself but making the need for it clearer. Still, the potential impact could be equally transformative.'[7] Thomas Friedman commented in *The New York Times*: 'Go figure: If this war doesn't inadvertently blow up the planet, it might inadvertently help sustain it.'[8] On the other hand (or edge), European nations fell back on coal and turned to increasing production from the USA and the Middle East for natural gas. Russia, which holds a critical share of global trade in key resources for the supply chain (e.g. 52 per cent of pig iron for steelmaking, 40 per cent of potash for fertilizer, 48 per cent of palladium for automotive catalytic converters), dangles the keys to sustainable future in front of its adversaries.

Primary energy grew by its largest amount in history in 2021, driven by a combination of emerging economies and a rebound in economic growth post-COVID. Oil demand increased, natural gas demand grew 5.3 per cent, above pre-pandemic levels, coal consumption grew 6 per cent, electricity by 6.2 per cent (of which wind and solar was 10.2 per cent and more than nuclear for the first time), nuclear grew by 4.2 per cent, and it was in that context that renewables (excluding hydro) grew by 15 per cent, more than 9 per cent on the year before and more than any other fuel in 2021. But decarbonization, resilience and productivity cannot be delivered by addition alone. We also need decentralization, to avoid transmission and distribution losses, and, more generally, we need to improve energy intensity, productivity, resource efficiency. We need to reduce.

- We need a mix of solar, wind, geothermal, air source heat, bioenergy, nuclear, hydrogen, and other clean technologies to provide lower-carbon energy in the system.
- We need to spend at least as much time on efficiency to ensure that time, money, and resources (whether conventional or renewable) are not wasted.

CHAPTER 6

Why We Need Energy Efficiency First

• • •

'EDGE is an acronym – it stands for efficient and decentralized generation of energy.'

Anon

- Energy efficiency is one of the largest and fastest sources of productivity and growth, as well as carbon emission reduction and energy security.
- Most energy is lost in the process of conversion, generation, transmission and distribution and more is lost at the point of use. Solving this problem cuts costs and carbon and improves resilience.

Resource efficiency is essential to, and for, sustainable development. Energy is one of the most essential resources we have – everyone and everything depends on it. It is highly valued, with profound economic and geopolitical implications and one of the largest investment markets in the world. It is associated with around 80 per cent of human-made greenhouse gas emissions, most of which comes from buildings, industry and transport. It is also a key interface between natural resources and the climate through the greenhouse gas emissions involved in converting it, generating it, transmitting it, distributing it using conventional energy, as well as the scarce time, money, resource and effort we put in to generating and adding alternative, lower-carbon energy.

And yet we waste most of it.

Solutions to this problem offer extraordinary productivity benefits, related resource efficiencies, cost efficiencies, energy security benefits, carbon emission reductions and improved energy security. Solutions are widely available today, based on existing technologies that are commercially proven and cost effective. Energy efficiency is sustainable because it is commercial. Energy efficiency should not come at the expense, or instead, of other actions to improve productivity and energy security, such as new, large-scale and innovative renewable energy capacity. But energy efficiency should come first because it reduces the size of the problem and because it can be delivered in the meantime.

Over 80 per cent of energy is oil, gas and coal, and that almost certainly will not change this decade.

Everything is going up. Demand for, and consumption of, conventional fuels is rising. At the same time, renewable energy is being added to the grid at increasing rates but from a lower base and with limits to its penetration in time and resources for development, planning, procurement of supply chain and construction. At this stage, utility-scale, grid-connected renewable energy is for the most part adding generation to the global energy system, not replacing fossil fuels at sufficient scale. To increase the challenge, replacing fossil fuels is the first enormous step. The second step is just as challenging because to 'electrify everything' we would need three to four times the amount of electricity that we generate today.

This is a problem because around 80 per cent of greenhouse gas emissions come from energy, and at this rate we only have eight years remaining in our carbon budget, that is, how much more greenhouse gas we can emit before we cause global temperatures to rise more than 1.5°C. The scientific consensus is that there are large and unpredictable consequences of breaching 1.5°C, some of which would be catastrophic for populations, nature, economies and security.

The problem with the problem is that most energy is wasted – for example, some 70 per cent of primary energy is lost in the USA before it gets to the point of use. In turn, 70 per cent of the end use goes to buildings, industry and transport.

So, what's the solution?

In theory, energy efficiency involves using less energy to achieve the same level of output. In practice, energy efficiency is achieved by energy saved in generating, supplying, distributing, or using energy.

So, using the numbers above, every unit of energy that you save at the point of use, say in providing energy services to buildings, industry and transport, is 2.3 units of energy supply that you don't need. In very round numbers – and recalling Warren Buffett's point that 'it is better to be approximately right than precisely wrong'[1] – of the 50 Gt of carbon we emit each year, 40 Gt is energy related, 30 Gt is in buildings, industry and transport, and at least 20 Gt is wasted. This wasted energy is the largest and cheapest source of greenhouse gas emissions reductions, as well as productivity and energy security. And we can achieve it this decade.

Where Does All the Wasted Energy Go?

Over two-thirds of energy is wasted before it gets to the point of use. Where does it go?

Before it gets to the point of generation, energy is wasted in the extraction and conversion process. For example, the refining process for gasoline is around 85 per cent efficient on average, meaning that some 15 per cent is lost in conversion. Fuel oil used in power plants has a 93 per cent refining efficiency. Researchers from Stanford University published a paper in March 2022 based on new evidence derived from cameras searching for methane leaks from wells, pipelines and storage facilities from the shale-oil and gas industry in the Permian Basin in Texas.[2] An estimated 9 per cent of the methane being produced was being lost to the atmosphere. Over a 20-year period, a tonne of methane released provides as much greenhouse gas-related warming as around 80 tonnes of CO_2. Unabated, these leaks could warm the climate three times more over the next 20 years than burning the gas sold. Russian gas is leaky too. The Rocky Mountain Institute estimates that Russian gas imported to Germany is twice as bad for the climate as LNG from the USA and three times as bad as gas from Qatar.

Now to generation. The law of the conservation of energy and the first law of thermodynamics both state that energy can neither be created nor destroyed, but can only change from one form to another. Primary energy is the energy that enters the economy in its original

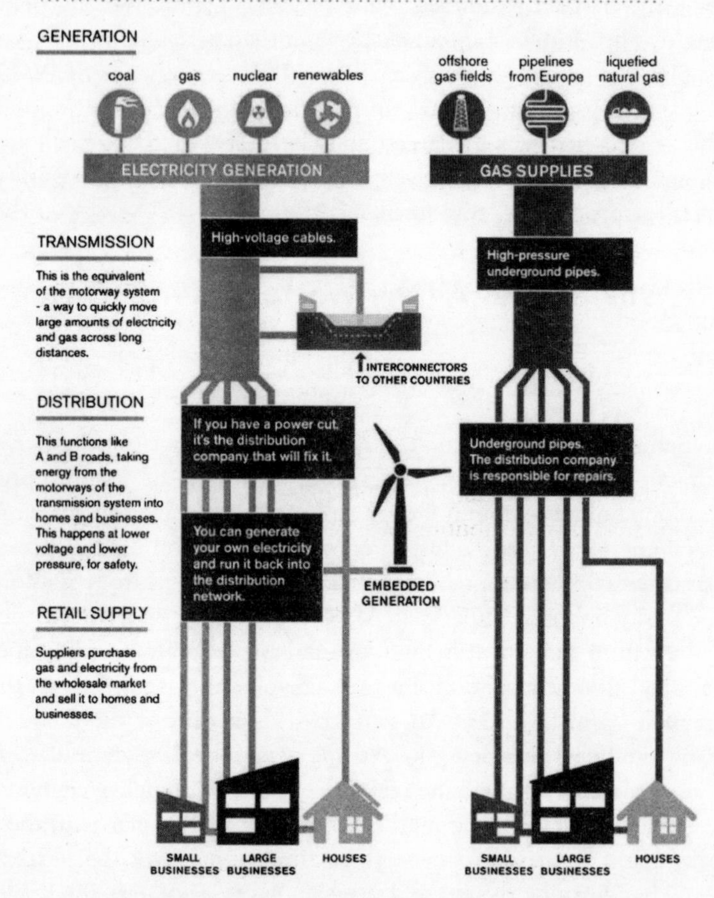

Figure 6.1 Britain's energy system and its components

Source: Ofgem (2019) 'State of the energy market'

form, e.g. gas, coal, oil, hydro, wind, solar, geothermal, bioenergy or nuclear. Useful energy that allows us to do work and perform services is 'exergy'. Energy services result.

The largest source of energy loss is heat. Gas is usually the largest source of electricity. Gas-fuelled power plants involve limits in terms of efficiency. Sometimes referred to as Carnot efficiency, thermodynamics dictates that half or more of the energy involved in the conversion of fossil fuels to electricity tends to be lost to heat. Heat from the power station, usually generated in a ratio of 1:1 alongside electricity, is difficult to use and it tends to get wasted. Some waste heat from electricity generation can be recycled, for instance by combined cycle gas turbines. But in practice, the limits of efficiency are typically in the range of up to 50–60 per cent. In the USA only 45 per cent of the natural gas consumed for electricity generation in 2019 was turned into electricity. In the UK, the thermal efficiency of the combined cycle gas fleet has averaged about 48 per cent. One of the main reasons that the heat is wasted is that gas power stations have tended to be built far away from the point of use in the context of a centralized energy system that typically produces energy for the grid. Given that gas tends to be the largest source of electricity generation for the grid in most countries, up to 50 per cent of energy can be lost out of the gate.[3]

Further losses, typically 5–10 per cent plus, can occur during the transmission and distribution process, taking the energy where it is

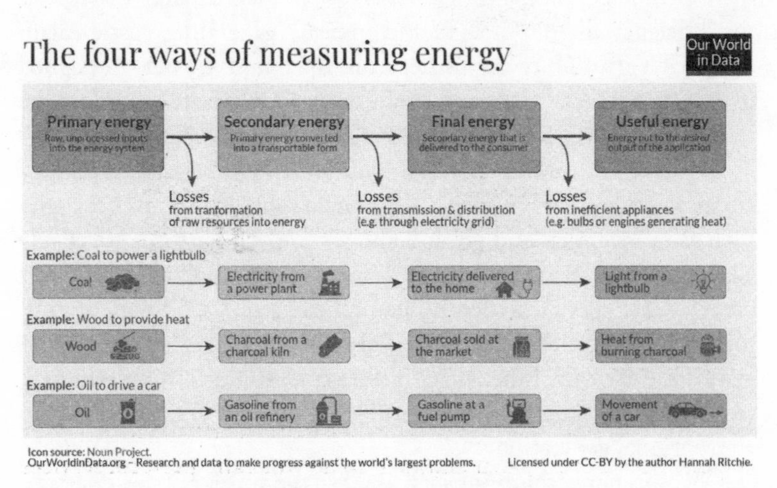

Figure 6.2 Four ways of measuring energy

Source: Noun Project and OurWorldinData.org licensed under CC by the author, https://ourworldindata.org/energy-definitions

needed. In the USA, the World Economic Forum estimates that some 70 per cent of energy is lost before it gets to the point of end use.[4] The Lawrence Livermore National Laboratory has been calculating and illustrating energy losses as energy flow charts or 'Sankey' diagrams since the 1970s and estimates that 67 per cent of total energy was wasted in the USA in 2021. A visual represention of this data is available at https://flowcharts.llnl.gov. Data from the International Energy Agency and the Lawrence Livermore National Laboratory related to waste in the energy economy is the opening argument of Tesla's Master Plan Part 3.[5] The biggest offender was transport at 78 per cent, compared to industry at 50 per cent.

What Can We Do About It?

Some of the problems can be addressed by capturing and recycling waste heat at the point of generation. Combined cycle gas turbines (CCGTs) can achieve efficiencies approximating 60 per cent. Carbon capture, utilization and storage (CCUS) promises to address the problem of carbon emissions, although as at the time of writing, there are no operational CCGTs with CCUS. CCUS may also reduce the efficiency of the CCGTs and increase generating costs. Allam Cycle CCGTs offer to go further from the outset, by being designed to recycle both the CO_2 and the heat and to operate at almost the same level of electrical efficiency as a conventional CCGT.

Most of the problems can be addressed by generating energy closer to where it is needed. While renewable energy technologies such as on-site solar and ground and air-sourced heat can effectively eliminate transmission and distribution losses, combined heat and power (CHP), even fuelled by natural gas, can slash losses by capturing and using the heat that is the by-product of the electricity generation process. Efficiencies of 80 per cent plus can be achieved, reducing greenhouse gas emissions by some 30 per cent compared to conventional generation.

On-site energy solutions such as CHP can be highly cost effective, typically reducing energy costs by 20 per cent if using the natural gas grid for fuel, or more if recycling waste gases, for instance from blast furnaces at steel mills or vent gases from wastewater treatment facilities. CHPs can be modified to run on hydrogen, when it is available, for reduced environmental impact. On-site solar, typically built on rooftops, has also become cost competitive with the grid. The European Union plans to make solar roofs mandatory for all new buildings later this decade. One major advantage of on-site or 'decentralized' generation is that it avoids the costs associated with the grid. Half of typical UK electricity bills have tended to be the wholesale price of energy, whereas the other half tends to consist of network, i.e. transmission and distribution system, costs and taxes. By avoiding the inefficiencies and costs of the centralized network, where practicable, costs and even carbon can be substantially reduced.

On-site generation also has the advantage of improving resilience and security of supply. Configured correctly, while the grid can be available as a backup, if the grid fails, which it has a habit of doing during severe weather events or times of supply constraint, then on-site generators keep the lights on. This can be critical for essential services such as hospitals, datacentres, industrial facilities and other key building users.

Transport is another key area of energy loss. Using oil to move a car involves even more energy loss than using gas to provide electricity to a building. Similar to the example of the gas-fired power station, a car's engine wastes heat in converting fuel to energy. The so-called 'well-to-wheel' efficiency of a car is estimated to be between 15 per cent and 30 per cent, compared to the equivalent 'wind-to-wheel' efficiency of electric vehicles, which is estimated to be 75–80 per cent.

Saving energy through more efficient use of energy or displacement for renewables and electric alternatives to transport significantly reduces demand for fossil fuels. Saving one unit of energy through more efficient supply is equivalent to removing the need for two or three units of energy in the first place. More efficient supply is not just lower carbon than business as usual, it is effectively negative carbon as it removes the need for energy in the system.

Another form of energy efficiency that is, effectively, negative energy or that produces 'negawatts' is demand-side energy conservation. This reduces demand for energy, typically by replacing less efficient mechanical and electrical equipment with more efficient equipment. The classic example is the LED lamp, which can reduce energy consumed by a conventional light by 60–90 per cent for the same quality of light, while lasting for 10–15 years.

Energy efficient technologies have experienced, and will continue to experience, very high growth rates. When I was developing energy efficiency projects for hotels and supermarkets a decade ago, efficient lighting technologies like LED were only 2 per cent of the world's lighting sales. Now they are over 60 per cent. But extraordinarily, relatively little has been done to date to take advantage of the energy costs and carbon savings that they offer. And it can be done. A great exemplar was a project that my firm worked on with GE Lighting for Santander Bank in 2015. It took around nine months to design and plan a project to replace every light bulb all their UK buildings (some 90,000 lights in over 800 buildings) with LED lights and, at the same time, to replace the heating, ventilation and air conditioning equipment and building management systems and controls. The project reduced related energy demand by about 50 per cent and resulted in enduring infrastructure solutions that would be fit for purpose for at least 10 years. All of this was a saving to Santander, not a cost. Consider today that less than a third of the lights in the UK's National Health Service (NHS), the public sector's largest energy user, have yet to be replaced. Changing the light bulbs would cost an estimated £350–400 million, but would save an estimated £1 billion over 10 years, so could be easily financed from savings.

Cooling, widely recognized to represent the largest and most cost-effective source of energy efficiency and greenhouse gas emission reductions today, is the next big thing – the Kigali Amendment to the Montreal Protocol is an international agreement to gradually reduce the consumption and production of hydrofluorocarbons (HFCs) in refrigeration – up to 2 billion new air conditioners could be sold in the decade to 2030.[6] The International Energy Agency predicts that the share of global electricity demand from air conditioners and fans will rise from around 10 per cent today to 20 per cent by 2050 as population grows and the planet warms. Demand for cooling is set to triple by 2050 and, if we do nothing, will be equivalent to all energy used by China and India today and require additional electricity capacity equivalent to the current capacity of the USA, the EU and Japan combined.

In the first five months of 2020, investments in new coal-fire power plants were two times faster than the same period in 2019. The number one driver of demand for their electricity is power for air conditioning. The International Energy Agency insists that with the right policies, we can double the efficiency of air conditioners, requiring fewer power plants, reducing emissions. All of the global growth in buildings' energy demand to 2050 can be fully met through energy and cooling efficiency improvements. Fifteen years ago, towards the beginning of my energy efficiency journey, variable speed drives that can save 60 per cent of energy by adjusting fan speeds to temperatures, which matters because 50 per cent of electricity is used in motors, were deployed in industry. Today, more efficient variable speed drives are available without reliance on rare earths. Building management systems, controls, insulation, battery storage and other techniques are increasingly pervasive, cost effective and quick to install.

Almost every type of building has proven to be a mine of energy efficiency opportunities. Take datacentres as an example. They usually consume 1–2 per cent of a nation's energy on electricity. Energy is so important to datacentres that they are measured in MW not square feet. They use energy to power racks of servers and cool the rooms that they occupy. Datacentres are so concerned about energy use that they use power usage efficiency (PUE) as a standard key performance indicator. Yet they cool the room rather than the rack. Most of the space being cooled is empty. Changing this relationship by cooling the racks rather than the room, as an example, could have a transformational impact on both space requirements and energy efficiency. Liquid cooling solutions that can do this are now being rolled out at a commercial scale. Many other building types offer similarly significant opportunities using existing proven technology. Waste heat and flue gas recovery from industrial facilities in the steel and cement industries, and wastewater treatment facilities are great examples. Indoor and outdoor lighting opportunities abound, particularly in hospitals and other public buildings.

Why?

The core merits and drivers of energy efficiency have been established and compelling for many years: cutting costs and carbon and improving resilience through the security of supply. Energy efficiency, put simply, can be cheaper, cleaner and more reliable. However, escalating energy costs, a sea change in concern about climate and sustainability for companies and governments seeking to achieve 2030 emissions reduction goals, and concerns about energy security, have driven a significant upswing in activity in the energy efficiency market. More recently, the war in Ukraine has blown the lid off each of the key drivers. Energy prices more than doubled in markets such as the UK. Decarbonization took a step backwards with the prioritization of energy security over carbon emission reductions. And while the big question was how to replace the 40 per cent of natural gas coming to Europe from Russia, the bigger question was and is: why are we still wasting most of it and what can we do about it?

The UK

An island nation, the UK imports about half of its energy and heats its homes on natural gas, with unusually low penetration of district energy or heat networks, compared to European peers. While natural gas boilers for household heat can be more efficient per se (c. 70 per cent) than centralized natural gas for power (48 per cent), it is nonetheless a problem from a cost, carbon and energy security perspective. The UK has relatively low levels of manufacturing and therefore lower levels of industrial demand than peers. All of this puts a lot of emphasis on displacing fossil fuels in households. However, the UK cannot entirely escape the laws of thermodynamics, and this leads to very large energy losses on the supply side and over 60 per cent of energy use is still outside the household.

On the supply side, the UK has a lot of potential for distributed and decentralized energy. Its centralized but distributed energy potential for offshore wind is the largest in Europe, thanks to the UK hosting 25 per cent of Europe's offshore wind resource, 10 GW today, with the ambition for up to 40 GW by 2030. On the decentralized side, there is an estimated 30 GW of CHP (six times today's installed capacity) and at least 10 GW of distributed solar, including on residential and commercial rooftops. On paper, this is enough energy, correctly balanced, to supply half the UK economy.

On the demand side, the UK Parliament's Business, Energy and Industrial Strategy Committee published a report in July 2019 entitled 'Energy efficiency: Building towards net zero', which stated that: 'The widespread deployment of energy efficiency measures across the UK's buildings will be a key pillar of any credible strategy to meet net zero greenhouse gas emissions by 2050, to tackle fuel poverty and cut energy bills. Energy efficiency investment also has the potential to unlock substantial, long-term economic returns.'[7]

The report proposed that total energy use could be reduced by an estimated 25 per cent by 2035 through cost-effective investments in energy efficiency (lighting, insulation, controls) and low- carbon heat – equivalent to the annual output of six Hinkley Point Cs (the newest nuclear power station procured in the UK). It found that reducing total energy use by 25 per cent by 2035 would result in average energy savings for consumers of roughly £270 per household per year, while at the same time sustaining between 66,000 to 86,000 new jobs annually across all UK regions. It estimated that a 'cost-effective' approach would require an estimated £85.2 billion investment but would deliver benefits (reduced energy use, reduced carbon emissions, improved air quality and comfort) totalling £92.7 billion – a net present value of £7.5 billion. Savings in the NHS, one of the UK's largest energy users and employers, of around £1.4 billion per annum were identified. It was pointed out that the impact of energy extends well beyond the walls of a hospital. The NHS is estimated to save £0.42 for every £1 spent on retrofitting fuel-poor homes. The present value of avoided harm to health was calculated at £4.1 billion in accordance with HM Treasury guidance. The report also found that energy efficiency can prevent expensive investments in generation, transmission and distribution infrastructure and reduce reliance on fuel imports – with a present value of avoided electricity network investment of £4.3 billion. As far as competitiveness was concerned, the report pointed out

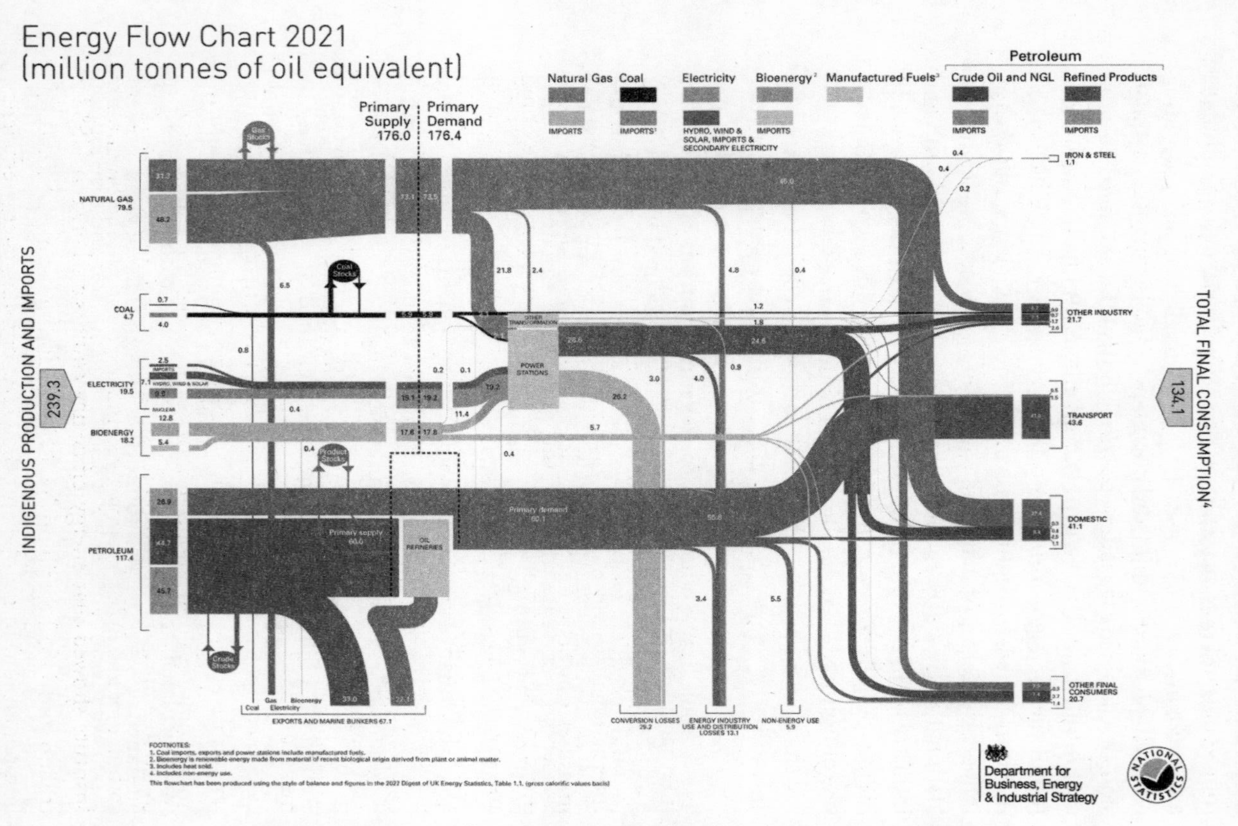

Figure 6.3 Energy flow chart 2021 (million tonnes of oil equivalent)

that the UK is a net exporter of insulation and energy efficiency retrofit goods and services. Low carbon, lower cost, more productive and resilient.

In 2022, faced with record energy prices following Russia's invasion of Ukraine, a broader economic crisis, and the need to plug a £50 billion budgetary gap, energy efficiency finally reached the top of the UK's government's priority action list. The Chancellor's Autumn Statement rightly called out energy efficiency and energy independence as of national importance. Indeed, it set a new 'national ambition' to reduce energy consumption in buildings and industry by 15 per cent by 2030. Doing so, it stated, could save £28 billion, or £450 for every household. An energy efficiency task force was announced.

Where Do People Stand?

International institutions have advocated energy efficiency to deliver both large-scale environmental and financial returns. The International Energy Agency has long held that energy efficiency measures can deliver US$1 of savings, on average, for every US$1 invested. This is certainly my experience, having been developing and investing in energy efficiency projects for nearly 15 years and in every major global market. A combination of on-site generation and demand reduction projects typically has a 'payback' period of six to seven years and a useful economic life of around twice that.

Considering social, economic and environmental benefit, the Copenhagen Consensus think tank estimates a return on investment of US$3 for every US$1 invested in doubling energy efficiency, compared with US$0.8 for every US$1 invested in doubling renewable energy. The United Nations Environment Programme's 'minimum ambition' model points to potential for annual savings by 2040 equivalent to 480 power stations, or 970 million tonnes of carbon dioxide. A 2015 report by Climate Works and the Fraunhofer Institute showed that energy efficiency could save between US$2.5 trillion and US$2.8 trillion by 2030, including up to US$150 billion per annum in Europe and the USA. Building energy efficiency,

Energy efficiency is imperative to tame total production costs that rose by 50 percent in recent months.

Potential energy reduction

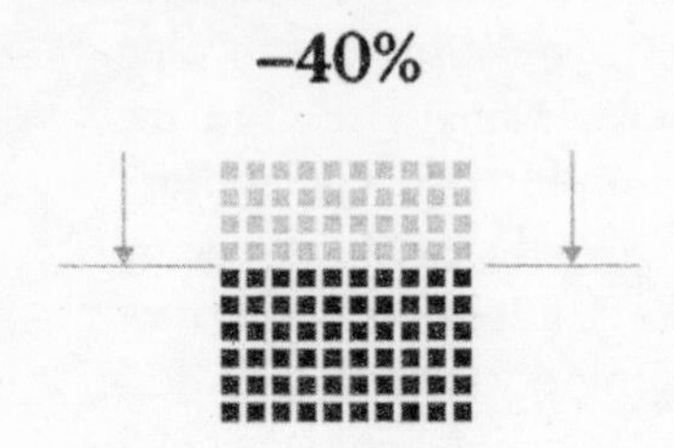

Heavy industry planned to reduce 50% of their energy and CO_2 footprint by 2030. Current utility prices make most of these business cases attractive much more quickly. Pulling forward high-impact cases could secure up to **40% energy-cost reduction** over the next 2–3 years.

Potential EBITDA[1] advantage

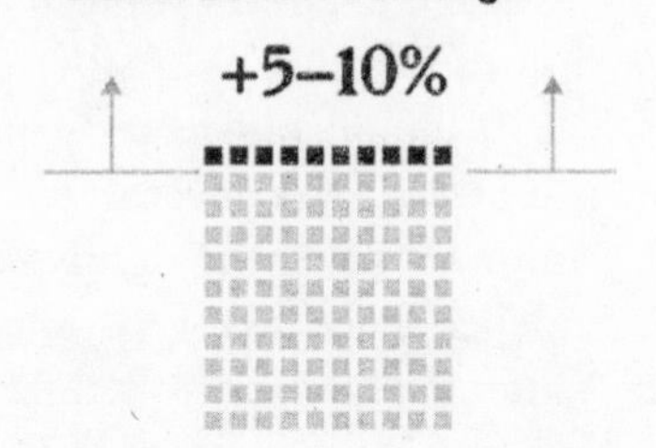

Companies taking bold action and performing a speed could make energy efficiency and supply competitive advantage worth **5–10% EBITDA margin** over sales, while abating their CO_2 footprint by more than 40%.

[1]Earnings before interest, taxes, depreciation, and amortization.

McKinsey
& Company

Figure 6.4 Energy efficiency is imperative to tame total production costs that rose by 50 per cent in the first half of 2022

Source: Exhibit 2 from 'The net-zero transition in the wake of the war in Ukraine: A detour, a derailment, or a different path?', May 2022, McKinsey & Company, www.mckinsey.com. https://www.mckinsey.com/capabilities/sustainability/our-insights/the-net-zero-transition-in-the-wake-of-the-war-in-ukraine-a-detour-a-derailment-or-a-different-path

according to the International Resource Panel, offers the highest value of resource savings of 15 groups of opportunities, at nearly US$700 billion, while at least four other energy efficiency opportunities made the list: iron and steel efficiency, transport efficiency, end-use steel efficiency and power plant efficiency.[8] Building energy efficiency was estimated to offer an average societal cost efficiency (calculated as the annualized cost of implementation divided by annual total resource benefits) of 0.5 compared, for example, to 1.2 for electric and hybrid vehicles. McKinsey's marginal abatement cost curves (see Figure 6.5) have always placed energy efficiency at the lowest end of the curve, amongst the highest impact at the lowest cost.

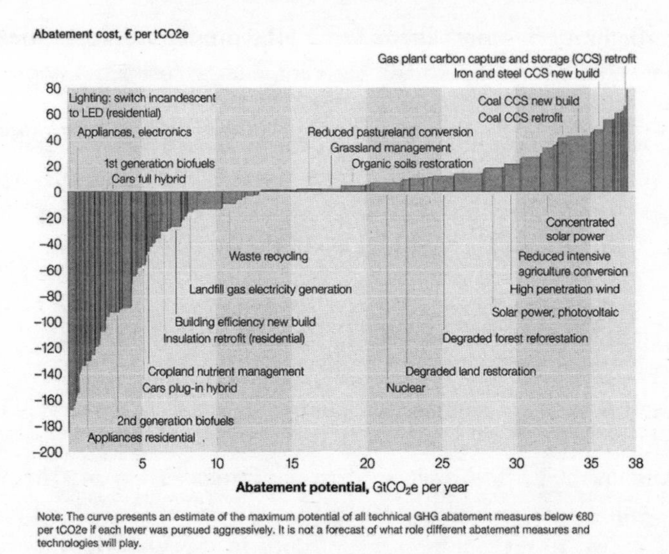

Figure 6.5 Abatement cost € per $tC0_2e$

Source: Exhibit from 'A revolutionary tool for cutting emissions, ten years on', April 2017, McKinsey & Company, www.mckinsey.com. Copyright © 2022 McKinsey & Company. All rights reserved. Reprinted by permission. https://www.mckinsey.com/about-us/new-at-mckinsey-blog/a-revolutionary-tool-for-cutting-emissions-ten-years-on

There are popular political tailwinds for energy efficiency, notwithstanding and partially because of the cost-of-living crisis exacerbated by high energy prices. Economically viable energy efficiency measures usually fail to attract sufficient or political public support. However, data from surveys now suggest that over 70 per cent of European citizens support regulation for improving home energy efficiency.[9] In the USA, 89 per cent of respondents to a March 2022 Gallup poll supported tax credits for home renewable-energy systems, 71 per cent were in favour of setting fuel-efficiency standards for cars, trucks and buses, and 61 per cent for tax incentives for the purchase of electric vehicles, among other policies.

As set out in a report by McKinsey in May 2022, in addition to driving the support for, and uptake of, energy efficiency, higher utility prices make the business case for hard-to-abate industry decarbonization more attractive.[10] High-impact, ready-to-deploy cases could secure up to 40 per cent energy-cost reductions and deliver significant additional earnings.

McKinsey's position is nothing new. McKinsey has been publishing marginal abatement cost curves since 2007. Figure 6.5 reproduces their update in 2017. While significant cost reductions have since been achieved in solar and wind, the message is simple.

What the Governments Are Saying

Governments have alighted on energy efficiency as a major potential source of cost-effective greenhouse gas emission reductions and, at least in theory, few more so than the European Union. The built environment is responsible for 40 per cent of energy consumption and 36 per cent of greenhouse gas emissions in the EU. The European Commission estimates that currently, over 75 per cent of the EU's building stock is energy inefficient. The 'energy efficiency first principle' is one of the key pillars of the EU's policies to help it meet its climate objectives as well as to reduce dependence on fossil fuels from abroad and to increase security of supply. It means taking the 'utmost account of cost-efficient energy efficiency measures' in shaping policy and making investment decisions. It aims to treat energy efficiency as a source of energy in its own right, in which the public and private sectors can invest, ahead of other more complex or costly energy sources. This includes giving priority to demand-side solutions whenever they are more cost effective than investments in energy infrastructure to meet policy objectives. The Commission drove home the principle in the recast Energy Efficiency Directive in 2021.

In 2020, the European Commission announced a 'Renovation Wave', alongside its 'Green Deal' and 'Fit for 55' initiatives. The Renovation Wave aims to double the rate of renovation for energy efficiency from 1 per cent of buildings per annum to 2 per cent per annum by 2030, involving some 35 million buildings in the 2020s. Three areas will receive special attention: tackling energy poverty and worst-performing buildings; public buildings, such as educational, healthcare, and administrative facilities; and decarbonization of heating and cooling. In line with a net 55 per cent overall greenhouse gas emission reduction target for 2030, the Commission expects the efforts of the Renovation Wave to reduce buildings' greenhouse gas emissions by 60 per cent, their final energy consumption by 14 per cent and energy consumption for heating and cooling by 18 per cent as compared to 2015. The Renovation Wave is expected to involve investment of €275 billion a year, around a third of which is expected as a direct investment from the Commission.

Then, on 18 May 2022, after being mandated by the European Council, the European Commission published a plan, 'REPowerEU', to phase out Russian gas as soon as possible. Among the numerous legislative initiatives endorsed, the European Commission proposed to make solar panels mandatory on all new buildings, to double the rate of deployment of heat pumps, and to integrate geothermal and solar thermal energy in modernized district and communal heating systems. The binding Energy Efficiency Target under the 'Fit for 55' package of European Green Deal legislation was enhanced from 9 per cent to 13 per cent. Energy savings had, for the first time, become a legal obligation of EU member states. The handle was cranked again in July when the European Parliament's four largest political groups reached consensus to demand a 14.5 per cent target, equivalent to a 42.5 per cent reduction in energy consumption by 2030. (The proposals to amend the energy efficiency directive implied a reduction of 40 per cent for final energy consumption and 42.5 per cent for primary energy consumption respectively when compared to the 2007 reference scenario projections for 2030.) In addition, the public sector was going to be required to lead by example. Public authorities in EU countries would be obliged to reduce energy consumption by at least 2 per cent a year, up from 1.5 per cent as previously instructed, to 'ensure that the public sector fulfils its exemplary role'.[11]

During its 7th Annual Global Conference on Energy Efficiency hosted in Sonderborg, Denmark in June 2022, the IEA implored policymakers to focus on energy efficiency policies. 'Energy efficiency is a critical solution to so many of the world's most urgent challenges,' said Fatih Birol, the IEA's executive director. 'Doubling the current global rate of energy intensity improvement to 4% a year' compared to today's policies would save 95 exajoules a year, the current annual energy consumption of China. In concrete terms, that would result in savings of 30 million barrels of oil per day, triple Russia's 2021 production. Additionally, the doubled ambition would cut fossil gas use by 650 billion cubic metres, four times the EU's imports from the Kremlin.

By July 2022, staring ahead at the prospect of winter energy shortages, Brussels cut to the chase. It asked EU member states to slash gas use by 15 per cent, an initially voluntary measure that would become compulsory if there was a 'drastic reduction', or shut off, of Russian gas. And in September it called for a 5 per cent reduction in electricity use. At the same time as reducing demand, it looked at decentralizing supply, bringing more to the point of use. In December 2022, the European Council slashed the maximum permitting procedure for roof-top solar power to three months, with the longer-term objective of mandating solar roofs for public sector, commercial and industrial buildings by the second half of the decade.

Several countries already have programmes that take advantage of the tax system to support energy efficiency investment by encouraging building owners to invest and receive tax refunds or repayments. The French MaPrimeRénov' offers a grant of up to €15,000 to all owners according to the type of energy efficiency work, with additional amounts for those in the low-income segment, in order to combat energy poverty. In addition, the Italian Superbonus provides a tax credit that is worth up to 110 per cent of the cost of the energy efficiency work undertaken and had already seen more than V25 billion spent under the programme as of March 2022. The German development bank KfW was reported to have invested nearly €37 billion in 2021 on improving the energy performance of existing and new buildings. However, major initiatives in the country were paused in early 2022 as funds had been exhausted. Other major OECD economies have provided more restrained levels of energy efficiency investment.

The USA has seen a minor increase in energy efficiency investment via the federal government weatherization programmes, increased to US$377 million in 2021, although a funds appropriation of more than US$800 million was requested from Congress in 2022. In June 2022, the White House authorized the use of the Defense Production Act (DPA) to accelerate domestic production of clean energy technologies, including solar panels, building insulation, heat pumps and electrolyzers. In July 2022, the re-positioned 'Inflation Reduction Act' provided US$9 billion of tax credits for low-income households to electrify their homes, and to buy heat pumps, rooftop solar panels and electric heaters. The Act incentivized the installation of efficiency upgrades and carbon capture in industrial sectors, potentially contributing up to 130 Mt Co2 of reductions. It also allocated an additional US$362 million to commercial energy efficiency tax deductions.

In some ways, while vastly more investment in energy efficiency is needed, we are starting from a higher base than many think. Although some 80 per cent of global energy investment still goes to adding energy through power and fuel supply, around 20 per cent goes to energy efficiency. Globally, spending on energy efficiency, electrification and renewables for end uses by sector was nearly US$500 billion in 2021, up from nearly US$400 billion per annum pre-pandemic. The largest share of investment related to buildings. A large driver of growth has been COVID-19 pandemic-related stimulus packages. By March 2022, governments worldwide had earmarked over US$710 billion in direct spending for long-term clean energy and sustainable recovery measures in COVID-19-related recovery plans. This is the largest ever clean energy recovery effort, 40 per cent higher than the green element of the fiscal stimulus disbursed after the 2008–10 global financial crisis, over US$450 billion towards energy efficiency in buildings and industry, developing mass transit and supporting the switch to electric vehicles. This amount comprises around 60 per cent of all recovery spending earmarked for sustainability. In July 2022, the German government announced plans to spend €177.5 billion on climate action, including €56.3 billion on supporting climate-friendly building renovation.

Why Isn't More Being Done Sooner and Faster?

If energy efficiency is cheaper than business as usual, if the costs of investing in LED lights and on-site energy generation can be covered by the savings in a limited period, generating substantial returns on investment, why does it not happen more? Why has the opportunity not been taken and why don't people put energy efficiency first? I face these questions every day because I raise money from investors to invest in energy efficiency projects every day. In fact, I have raised and put to work over US$2 billion for profitable energy efficiency projects and companies over the last decade. While there has been some exciting development in technology from a performance, cost and availability perspective over the last decade, I often say that in most respects we were saying that the same thing could be done a decade ago. The only difference now is that there are at least 2 billion reasons we're right. So, the answer?

Some of the headwinds will come down to implementation. First, rising construction costs and high inflation will reduce the impact of higher levels of investment as the money will not go as far. According to the IEA, buildings efficiency support disbursed through direct subsidies and tax rebates is particularly at risk of under-delivery, as complex funding application processes hold back wider uptake. Global supply chain constraints, labour shortages and price hikes are also delaying construction projects, causing budgets to overrun or contractors to shelve projects in several sectors, from housing retrofits to large mass-transit infrastructure.

But possibly the most candid reason is one that I was given by a veteran practitioner in 2006 when I got started. Invoking an atavistic chauvinism in need of being stamped out, he explained, 'Real men,' he said, 'build power plants.' For me, no explanation since has ever summed up so well the rigid captivation with the supply side of the energy sector, nor the reality of incentives in the energy supply chain. Utility business models are built on providing centralized energy services to a centralized grid system. Their embedded supply chains consist of equipment manufacturers building and selling large-scale generation equipment and fuel suppliers selling oil and gas and coal. The utilities sell to either privately or state-owned and controlled national grid system operators, who in turn supply to local distribution network operators. Margins, market incentives and subsidies are intricately layered into every step in the supply chain. So embedded are the supply chains that national champions, sometimes supported by the state, can even find themselves in partnership with competitors, not just at corporate level but at state level, where the nation states are in physical or financial conflict. Good examples are Shell, E.ON, Uniper and Engie and their business, variously, with Gazprom and Novatek.

In contrast with the well-organized and centralized mainstream energy system, decentralized energy markets are by nature highly diverse and fragmented, and involve-smaller scale systems delivered by and for a very wide range of counterparties. Utilities tend to serve the grid, which is a conduit, not the end customer. By contrast, decentralized energy and energy efficiency markets serve the end customer directly. The end customer in the non-residential markets alone consists of millions of corporate, industrial and public sector counterparties. Even solving for the 80:20 rule, where 80 per cent of the value is in the top 20 per cent of the market, this still involves tens or hundreds of client counterparties in any market. These counterparties, unlike the grid, have feelings. They are highly concerned about risks to availability, costs and, often, meeting carbon and other environmental goals.

The arrangement of a market to serve the end customer is a matter of evolution rather than revolution and does not lend itself easily to the 'one-to-many' route to market that characterizes the handful of large companies and service providers that dominate the centralized energy system. Sectors such as datacentres and hospitals can represent very large proportions of national energy demand. Datacentres in Denmark and Ireland have been responsible for upwards of 15 per cent and 30 per cent respectively of national or municipal energy demand. Hospital systems such the National Health Service in the UK can represent 2–4 per cent plus of national demand. So too public sector departments such as the Ministry of Defence or government-backed transport systems such as Transport for London. However, there are several parties involved in decision making in any of these sectors, making decisions on behalf of a number of often entirely independent stakeholders. Decision makers includes finance directors and engineering departments, which very often speak entirely different professional languages. What, after all, is a single line diagram?

Smaller-scale projects often take the same amount of effort as larger projects. Even where they make economic sense and stand on their own feet financially, they can struggle to attract as much or even any investment because they lack economy of scale for the investor. Smaller projects attract smaller developers and advisers, which can have less capacity and fewer resources. Smaller projects can get to scale through replication, but this can be hard to achieve with bespoke requirements of individual buildings or facilities. Projects are required to be designed from feasibility stage and then approved to move forward to conceptual and detailed design. They require a diverse set of front-line skills, a multi-disciplinary approach. They are subject to procurement and accounting challenges, particularly in the public sector, which can result in months or years of delay. The project development and sales cycles of projects in the public and private sectors very often run into years, during which time decision makers change jobs or retire. But these problems are solvable with persistence and long-term commitment. Indeed, the solutions to these challenges create the competitive advantage for the best actors in the market, who can solve problems for the source of energy demand, the end customer.

Solving Problems for Customers

It is worth pausing here for a moment, at the point of use, at the customer. The customer is often blamed for the problem of energy waste. This can be unfair. As we have seen, some two-thirds of energy is lost before it gets to the point of use in the first place. Most of the battle has been lost before the consumer's fight even starts. There is much more to do at the point of use – and indeed shareholders and taxpayers should be asking very tough questions of their companies and governments if they are wasting energy and money – but it can and must be commercially viable, and here the playing field can be levelled. The cost-of-living crisis exacerbated by energy and food price increases in the UK and Europe in 2022 forced the decision on allocation of household incomes between food and fuel. This decision becomes harder into the winter and heating and hunger create two essential and competing demands. Energy measurement, management, validation, certification, rating and labelling schemes have been introduced in several countries including the UK (Building Research Establishment Environmental Assessment Method, or 'BREEAM'), the USA (Leadership in Energy and Environmental Design, or 'LEED', and 'Energy Star' – a programme run by the US

Environmental Protection Agency and the US Department of Energy that promotes energy efficiency by providing information on the energy consumption of products and devices and assigns labels to qualifying products), Japan (Top Runner) and Australia (National Australian Built Environment Rating System, or 'NABERS').

Government regulations can limit energy waste all the way through from the point of construction to long-term operations. For instance, the Netherlands introduced laws in 2013 that limit carbon emissions per square metre of construction. The UK government introduced the Minimum Energy Efficiency Standards in 2015, whereby it was deemed unlawful to rent out commercial and residential properties with an Energy Performance Certificate (EPC) rating below a certain threshold. Minimum energy performance standards (MEPS) limiting the amount of energy that appliances can use were introduced early on in California, Australia, New Zealand, and Brazil. Government regulation is also the most straightforward way of dismantling the often-cited Jevons Paradox, derived from an observation by William Stanley Jevons in 1865, that if the UK used coal more efficiently, then it would end up wanting more of it, implying that energy efficiency leads to an increase in total demand. In any case, over 150 years later, resource constraints and global problems associated with energy cost, energy security, carbon emissions and pollution have turned such arguments about 'rebound' on their head. In New York, Local Law 97 requires most buildings over 25,000 square feet to meet energy efficiency and greenhouse gas emission limits by 2024 or face fines.

Shifting Gears

In the past, decentralized energy and energy efficiency projects have not been considered, for want of a better phrase, 'sexy'. Energy costs were a factor but rarely one of the largest overhead items vexing finance directors. Energy security was, largely, taken for granted, as the grid was assumed to be available whenever needed. Despite a combination of high-level policy and lip service, few cared about or acted on carbon. Over the last few years, these factors started to change. Energy prices escalated. Under-investment in grid infrastructure and disruption, particularly in the USA, from severe weather events tested confidence in reliability of supply in the grid, particularly when blackouts hit New York, Texas, Louisiana, and other markets. Then carbon and ESG became a 'thing', in fact quite a big thing, surrounded by a wave of measures and new government regulations to monitor, measure and report. But the lid was blown off by Ukraine. Costs exploded, doubling and more. Energy security became the hottest topic in Europe as supplies from Russia were constrained. And the carbon and ESG debate came to the fore, as Europe had to lean back on whatever sources of natural gas and oil were available. Doubling down on renewables as an answer to energy security and climate became a mantra for the UK and European governments, but the mantra also revealed the shortcomings of the strategy to deliver solutions at any meaningful scale well within the decade. Time runs out.

Unless we focus at least as much on reduction and efficiency as we do on addition and capacity, we risk failing in all our efforts to mitigate climate change. We will also have lost productivity gains crucial for prosperity and sustainable growth, as well as the opportunity to improve energy security and resilience. Because climate change is a global problem that requires global solutions, we also must think and act beyond our borders. The environment and climate operate, of course, beyond territorial borders. As does economics. And society.

Consider population for example. The global population is predicted to hit 10 billion in 2050, leading to increased energy demand for both renewable and traditional energy sources. Many predictions state that global energy demand will increase by nearly 50 per cent in the next few decades.[12] According to the IEA's Net Zero by 2050 Pathway, without significant gains in energy efficiency the growth of energy demand due to population increases alone would present a substantial challenge in transitioning to alternative, lower-carbon sources of electricity. Without fundamental consumption changes that include a push towards maximizing energy efficiency, final energy consumption in 2050 could be 90 per cent above what is required to achieve the net zero pathway. We would otherwise be asking renewables to increase fivefold in scale, from 20 per cent of the global energy system to 100 per cent of the global energy system, then to scale up to four times more energy because of electrification and then up to 50–90 per cent more because of population and the advance of developing societies. Even if this could happen, it will take longer than we have. Instead, we need to reduce energy use, not just add to it. According to the IEA's March 2021 report, 'How energy efficiency will power net zero climate goals', 'maintaining global growth and supporting development in emerging economies implies a sharp rise in consumption habits. Meeting this need requires a transformation of the existing energy system'.[13]

Addressing the Crises of Climate and Energy Security

Time is also a key factor, as the challenge gets larger as time goes on with population and consumption growth. The European Commission's focus on energy efficiency recognizes that its carbon emission reduction goals for 2030 cannot be achieved on the supply side alone and

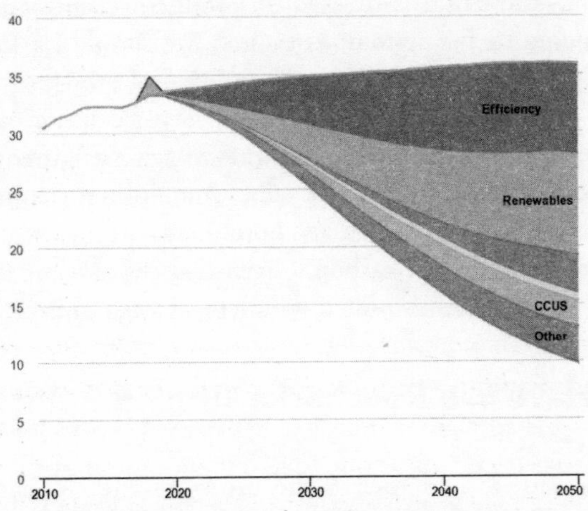

Figure 6.6 CO_2 emissions reductions by measure

Source: IEA (2022) 'CO_2 emissions reductions by measure in the Sustainable Development Scenario relative to the Stated Policies Scenario, 2010–2050', https://www.iea.org/data-and-statistics/charts/co2-emissions-reductions-by-measure-in-the-sustainable-development-scenario-relative-to-the-stated-policies-scenario-2010-2050

without action to reduce demand, 'reducing the size of the cake'. According to the IEA's Sustainable Development Scenario, energy efficiency represents more than 40 per cent of the emissions abatement needed by 2040.[14]

Putting many of the pieces together, the 2022 Sixth Assessment Report on Mitigation of Climate Change by the UN IPCC heavily emphasizes the long-term importance of these measures in relation to 2050 goals, finding that demand-side measures and innovative end-use service provisions can reduce global greenhouse gas emissions by 40-70% by 2050, requiring transitions in fossil fuel use, low-emission energy sources, and energy efficiency.[15]

In parallel to the climate crisis, the energy security crisis demands urgent action on energy efficiency. By June 2022, the flow of natural gas to Europe from Russia had reduced by up to 60 per cent. Concerns were heightened of a complete shutdown. Aside from hitting the needs of the present, it created problems for the future as gas storage facilities struggled to fill up.

Without more efficiency, we are pouring fresh water into a broken vessel. While we are on analogies, we are shovelling water with a pitchfork. The world is focused on producing, adding, more and cleaner energy to the system, but surely we should fix the system, where it is broken or leaking, to ensure that it is fit for purpose? If you were refurbishing your home with new furniture, would you not clean the home first? We risk failing to achieve either climate or energy security goals and, at the same time, losing the opportunity to succeed in making economies more productive, prosperous, and resilient, as well as lower carbon. Given that the clock is ticking, we will need more efficiency now if we are to succeed in time.

Have We Reached a Turning Point?

The propensity to add fuel to the fire is, essentially, contributing to a potentially zero-sum game in the fight against climate change. But there are other energy fights raging, about security and cost. Russia's invasion of Ukraine reduced supplies of crucial oil and gas to Europe. Supply constraints, in conjunction with the effect of sanctions, increased energy prices, creating a high margin and more profitable business for Russia but a commercial and cost-of-living crisis for Europe. The immediate responses were a dash for alternative sources of gas and an escalation in ambition to 'double down' on building new renewable energy capacity. As the months wore on, neither solution tackled the problem. Renewable capacity at the scale to have any meaningful impact on the problem would take decades. Alternative sources of gas are scarce and come at a cost. By June 2022, the European Union's Energy Commissioner, Kadri Simson, said at a conference in Denmark, 'We will not find sufficient volumes of gas globally to replace Russian gas ... Even if Ukraine prevails, we should not end in the same position where we're so dependent on fossil fuels.'[16] Energy savings, says the EU, are necessary to mitigate risks to supply security. We cannot fix energy security, energy cost or carbon emissions on the supply side alone.

Will we learn from, or repeat, recent history, or take a different course?

Russia invaded Crimea in 2014. At the same time, there was a campaign for energy efficiency. The European Commission cited that every 1 per cent of energy saved means a 2.6 per cent reduction in gas imports. EU energy commissioner Günther Oettinger pushed for an energy efficiency target of 30 per cent below a 2007 baseline, while NGOs and think tanks called for a 40 per cent target. The European Commission itself calculated that a 40 per cent target would raise GDP four times more than 30 per cent. The European Council (member states) settled on a number of 27 per cent. However, since 2014, Europe's dependency on gas imports has grown. According to Niels Fuglsang, the Danish MEP leading the revision of the Energy Efficiency Directive in the European Parliament: 'It is not just about business, it is about security policy. I think it will change this time because the Ukraine situation is the biggest security threat to Europe since the Cold War.'[17]

- 80 per cent plus of energy is oil, gas and coal.
- 80 per cent of human-made greenhouse gas emissions come from energy.
- 80 per cent of energy investment goes into the supply side.
- 70 per cent of energy is used in buildings, industry and transport.
- 70 per cent of primary energy is lost before it gets to the point of use in the USA and other markets.
- 20–30 per cent of energy can be lost at the point of use.
- Productivity is bound to be limited if we waste most of one of the most valuable and essential inputs into the economy.
- Energy efficiency is one of the largest and fastest sources of productivity and growth.

CHAPTER 7

The Hard Truths About Natural Capital

• • •

'Our civilization is constitutionally incapable of reversing the annihilation of natural capital, or even slowing it down. Get used to that. When we really understand that, the project of reconceiving civilization itself will gain powerful impetus.'

Charles Eisenstein

- Like energy, we waste nature. In fact, we are laying nature to waste.
- This is going to prove very costly. It will hurt the climate, cost trillions, and leave societies insecure, hungry and at war.

Franklin D Roosevelt, 32nd President of the USA, once said: 'A nation that destroys it soil destroys itself. The forests are the "lungs" of our land, purifying the air and giving fresh strength to our people.'[1] But the problem is bigger than that. It is no longer a national problem, it is a global one.

According to the Goldman Sachs 'Carbonomics' report, in order to limit global temperature rise to 1.5°C, agriculture and forests need not only to reduce their contribution to emissions from 5 $GtC0_2$ to 0 $GtC0_2$ by 2030, but also then to become carbon positive and remove or absorb 3 $GtC0_2$ by 2040. Forests are our largest carbon sink. But we are laying them to waste.[2]

Globally, forests now represent some 30 per cent of land cover, but our 3 trillion trees today are around half the number they were at the beginning of the twentieth century. We have cut down around 1.5 million square kilometres of them since 1990, including around 10 per cent of the Amazon rainforest. Between 2014 and 2018, on average an area the size of the UK was lost each year, 90 per cent of which was tropical forest. Around 80 per cent of deforestation is clearing for crops or livestock and the rest for timber logging, mining, urbanization and infrastructure. Animal agriculture is responsible for 91 per cent of Amazonian deforestation. The Amazon hosts half of the tropics' undisturbed forests and absorbs 1.5 billion tonnes of CO_2 each year, roughly 4 per cent of emissions from fossil fuels. As *The Economist* put it, 'If rainforests were Earth's respiratory system, the Amazon would be a full lung.'[3] Yet an area of the Amazon five times the size of London is being cut down or burned every year, destroying trees that absorb CO_2 but also releasing their stored carbon back into the atmosphere. The Brazilian Amazon is now a net carbon emitter. Over the last 20 years, it has lost 350,000 square kilometres and emitted 13 per cent more CO_2 than it has absorbed. A new high-resolution map produced in February 2022 showed that fires caused a third of global forest loss between 2001 and 2019, particularly in the tropical primary forests of Africa and Latin America. Severe wildfires in Australia, California and Brazil in recent years have been high-profile examples of threats of forest fires.[4] While some of this loss would have been caused by temporary slash-and-burn farming, some would have been from longer-term deforestation and some from natural wildfires ignited by lightning (some major forest fires were also a result of very unnatural causes: sparks from electrical distribution lines that could have been prevented by burying power lines) but all emitting CO_2. Fire seasons are becoming more widespread and 'fire weather', associated with hotter and drier conditions, has been attributed to climate change in the IPCC Sixth Assessment Report.[5]

Forests are a vital part of a wider category of terrestrial systems, including forestlands, peatlands, wetlands, grasslands, mangroves, tidal salt marshes, farmland and rangeland. The world's largest area of tropical peatlands, made up of partially decomposed trees and plants, is in the Congo. It holds the equivalent of around three years of global emissions from fossil fuels. However, in July 2022, 27 oil-exploration blocks were auctioned encompassing up to 1 million hectares of peatland and up to 11 million hectares of rainforest, which sit on top of an estimated 16 billion barrels of oil worth around US$650 billion.

Collectively, terrestrial systems hold 3.3 trillion tonnes of CO_2 in and above ground, around four times more CO_2 than is in the atmosphere.[6] Degradation, development, conversion or loss of these ecosystems increases CO_2 emissions. A loss of 10 per cent of these terrestrial systems (noting the rate at which we are losing the Amazon) could increase CO_2 by as much as 100 parts per million. This implies very severe levels of global temperature rise. These terrestrial systems, or natural infrastructure assets, in turn depend on other natural living things, which may not be obviously associated with the climate. 75 per cent of bird species and 68 per cent of mammal species live in forests. Pollinators, wildlife corridors, habitats, bioregions, wildlife migration, grazing ecology, seagrasses, beavers and more generally the birds, reptiles, rodents, mammals and insects that occupy terrestrial systems are essential to their survival; if they fail, the systems fail. Biodiversity and survival are inseparable.

Forests also connect directly to the water cycle and deforestation can cause devastating disruptions. A stark example is India, where rampant deforestation in the Western Ghats, which was at one stage a picturesque landscape of valleys, gorges and virgin forests and home to more than 30 per cent of India's species, is causing recurring flood in southern India. Large dams on major rivers offer renewable energy but can also bring deforestation and destruction of ecosystems. Areas that were not previously considered flood-prone, such as Maharashtra, Karnataka, and Kerala, are now at risk. Stronger and more erratic rainfall in recent years, attributed to climate change, creates recurrent floods in low-lying areas, with risks increasing in line with population growth. In June 2022, floods in Assam, following rainfall described by residents and authorities as being of 'biblical proportions', displaced more than 4.7 million people, destroying homes, and cutting off access to food and water. A local environmental science professor, while acknowledging the increase in frequency in rains, observed: 'Before linking it entirely to climate change, we need to take into account human-related factors like deforestation.' Felling trees near rivers removes roots that can hold large amounts of water.[7]

The natural carbon cycle story continues from the land into the oceans. While forests, plants and phytoplankton absorb CO_2 and convert it to oxygen and carbohydrates, around 25 per cent of CO_2 emissions are absorbed into oceans. Some of this is converted into fish and other marine life. The rest is converted to carbonic acid and kills sea life. Oceans are the largest carbon sink on Earth, containing 12 times more CO_2 than land and 45 times more than the atmosphere. Oceans absorb heat and around a third of the CO_2 released each year from human activity. Some 93 per cent of increased atmospheric heating has been absorbed by the oceans, resulting in warming ocean temperatures and acidification, which makes it harder for some marine life to live and, in turn, for the oceans to sequester carbon. This is pushing the ability of the oceans to continue acting as carbon and heat sinks to the limits, to the edge. In April 2023, a new study into the Earth energy imbalance (EEI) identified an unprecedented rate of warming of oceans, adding to fears of a negative feedback loop in fuelling a strong El Nino Southern Oscillation, which could further drive warming.[8]

The first to consume CO_2 are phytoplankton, microscopic plants, on the water's surface. Obtaining their energy through photosynthesis, they produce the foundation of the ocean's food web as a food source for all marine life. Carbon is cycled throughout life in the oceans as smaller life forms are consumed by bigger ones. All marine species contain carbon, but phytoplankton stand out, holding up to 2.4 billion tonnes of carbon, almost the sum of the CO_2 sequestered by all the trees, grasses and other land-based plants combined. Phytoplankton are essential to the food web but also to the carbon cycle because when they sink or die, they transport carbon to the deep ocean and, in so doing, are the main vehicle for the long-term removal of carbon from the oceans and the atmosphere. They are responsible for around half of global photosynthetic activity and at least half of the oxygen production, despite constituting only about 1 per cent of the global plant biomass. The effect of global warming on phytoplankton is the subject of ongoing research, but attention is being paid to factors such as ocean acidification, availability of nutrients, B vitamin deficiencies and warming ocean temperatures, which may increase species richness and bloom, but may also shift distribution towards the Earth's poles, which has the potential to disrupt ecosystems and the sustainability of fisheries. A shift in location also has the potential to reduce phytoplankton's ability to store carbon and could disrupt the carbon cycle. Oceans cover 70 per cent of the Earth and are essential to the carbon cycle, but also to feed billions of people. They deserve protection.

Rather than protecting oceans, we are polluting them. Coastal waters contain plastic pollution from land, shipping, drilling, fishing and direct dumping, together with industrial chemicals, petroleum and fracking waste, agricultural runoff, pesticides, pharmaceuticals, raw and treated sewage, heavy metals, radioactive waste, and street waste from urban runoff. The devastation to the environment in the ocean and at the coastline is clear and visible to the naked eye. Up to 12 million tonnes a year of plastic waste is estimated to enter the oceans. Sometimes, the floating debris gets caught in slow turning whirlpools to create visible monuments to pollution. The best known of these is the Great Pacific Garbage Patch, which is a mass approaching the size of Alaska that spins clockwise in the Pacific Ocean. It is one result of the production, incineration and disposal of plastic that contributed more than 2 billion tonnes of greenhouse gases to the atmosphere in 2020, equivalent to the emissions from nearly 500 coal-fired power plants. Greenhouse gas emissions from petroleum-derived plastics are on track to grow to 6.5 billion tonnes by 2050.

Society depends on the land and water for all its food, for survival. Around 34 per cent of total greenhouse gas emissions are caused by the global food system, including production, transport, processing, packaging, storage, retail, consumption and waste. Food that gets spilled or spoilt before harvesting – or when being stored, packed, or transported – is known as food loss. Food loss can be caused by weather or insects, or by high temperatures and poor storage. Some food can be lost as it is discarded in the distribution process for cosmetic reasons, imperfect vegetables for example.

Around a third of food produced for humans is lost before it reaches the consumer (like clothes in the fashion industry). Up to 40 per cent of the food supply in the USA is wasted each year, worth some US$161 billion. (The amount of food waste is estimated to weigh some 133 billion pounds, equivalent to 10 million elephants.[9]) Around 90 per cent of food waste in the USA ends up in landfill and produces methane when it rots. Food waste contributes 9 per cent of global greenhouse gas emissions, or 12 per cent including landfill emissions. These are solvable problems. China has sought to reduce food waste under its 'Clean Plate Campaign'. Efficiency and the reduction of spoilage during harvest, processing and distribution can make a big difference, as can cold storage systems. Redistribution of food has an impact – food banks reduce some 10.5 million tonnes of greenhouse gas emissions a year. Food and agricultural waste can even be recycled as energy.

War

Land is essential to any definition of sustainability and any plans for net zero and climate change mitigation. However depletive humanity has been of its natural capital in the ordinary course, it is made worse by conflict. Supply shortages and price increases in agricultural markets following Russia's invasion of Ukraine were feared to lead to conversion of more land to agricultural production, increasing deforestation rates and agricultural emissions. This seems to be playing out. The war disrupted the global timber trade. Sanctions curbed supplies from Russia, the world's largest softwood exporter, as well as Belarus and Ukraine. The three countries together represented a quarter of 2021 worldwide timber trade and around 10 per cent of Europe's demand. The shortfall of timber, like the shortfall of energy, is addressed by relaxing environmental protections to increase production. Ukraine started by lifting limits on logging in protected forests, boosting wartime export earnings. Estonia, Finland, and the USA also aimed to increase logging volumes. This added to an already challenging backstory. An outbreak of spruce bark beetle had led to emergency harvesting in central European and Alaskan forests. Intensive tree felling in Finland transformed its forests from a carbon sink to a source of emissions for the first time in 2022.

Within only six months of the pledge by 100 world leaders at the COP26 summit to end deforestation by 2030, the situation appeared to be going backwards. Back in Ukraine, forests faced uncontrolled wildfires, particularly in the frontline Donbas region. As a spokesperson from the Ukrainian Nature Conservation Group put it: 'There is no one to fight them.'[10] The private sector was not fighting. An analysis commissioned by the UN Climate Change High-Level Champions found that out of 148 major food, land and agriculture companies committed to net zero emissions by 2050, only nine were making progress to end deforestation and over 90 per cent risked missing their net zero commitments because of a lack of action on deforestation.

Law

The complaint threatens to turn legal. Campaigners argue that regulators should hold banks and pension funds accountable under money laundering legislation for profiting from illegal logging, fishing, waste trafficking and wildlife trade. These 'nature crimes', largely committed by producers of food such as soya and beef in Brazil and palm oil in Indonesia, are estimated to generate US$280 billion a year in revenues and to result in US$30 billion a year in foregone taxes. Mooted 'ecocide' legislation raises the prospects of lawsuits against many companies.

An early example was the case brought by a Peruvian mountain guide, Saul Luciano Lliuya, against German utility RWE, based 10,500 kilometres away in Essen and the second largest emitter in Europe.[11] He claimed that RWE had a role in melting the Palcaraju glacier, which posed a danger to his home and livelihood, and he demanded compensation for the harm it caused him. The case, brought in 2014, was originally rejected, but then was picked up by an appeals court in Hamm. In the ongoing case backed by Germanwatch, a campaign group, Luciano Lliuya wants RWE to contribute to the city's flood defences pro rata to its contribution to the greenhouse gas emissions that are held partially responsible for the glacier melting. The damages sought are 0.47 per cent of the cost of a proposed dam and improved flood barriers for the city, equivalent to €17,000. The figure corresponds with RWE's estimated greenhouse gas emissions in a 2014 report on 'carbon majors' by the Climate Accountability Institute. According to the study, RWE was responsible for 0.47 per cent of the world's carbon dioxide and methane emissions between 1750 and 2010. The court, however, will only consider RWE's emissions from 1958 onwards, which is 'the point from which a company could reasonably know that climate change is a problem and do something about it', according to Germanwatch. RWE says it's wrong to single out a single company over a global phenomenon. A company spokesman summarized RWE's defence as follows: 'Individual emitters are not liable for universally rooted and globally effective processes like climate change ... It is judicially impossible to relate specific or individual consequences of climate change to a single person.' The court hearing in the Andean city of Huaraz took place in May 2022. Judges of the Higher Regional Court of Hamm, legal advisors and experts travelled to Peru to check whether the plaintiff's house is actually threatened by a possible flood wave from the Palcacocha glacial lake above the city. A decision has not yet been made.

Then, in June 2022, a case was brought by the Australian Conservation Foundation (ACF) against Woodside Energy's US$12 billion project to develop an 11 trillion cubic feet natural gas field, Scarborough, off the west coast of Australia, which would emit the equivalent of up to 1.37 billion tonnes of CO_2 over 20 years, or more than Australia's total annual carbon emissions of around 500 million tonnes. ACF argues that the emissions will directly contribute to the destruction of the Great Barrier Reef, 3,000 kilometres away on Australia's east coast. It used 'attribution science' to prove a link between greenhouse gas emissions, rising global temperatures and coral bleaching that kills coral reefs. Scarborough (which Woodside claims would produce gas for power generation at half the emissions of coal generation, and had been the subject of rigorous environmental assessments by a range of regulators) has never been approved under Australia's federal environment protection law – the Environment Protection and Biodiversity Conservation Act 1999 (Cth) (EPBC Act) – because offshore gas and oil projects are assessed in a streamlined process by the National Offshore Petroleum Safety and Environmental Management Authority (NOPSEMA), which approved Scarborough, and exempt from the EPBC Act's operation. However, that exemption does not apply if an offshore project is likely to have a significant impact on the World or National Heritage values of the Great Barrier Reef. ACF's case relies partly on this application of federal law and an argument based on attribution science could break new ground. But the ice was broken in May 2021, when Royal Dutch Shell was ordered by a case brought to the District Court of the Hague (by Friends of the Earth Netherlands and six other environmental and human rights groups, together with 1,700 Dutch citizens) to cut its carbon emissions by a net 45 per cent by 2030, compared to 2019 levels, to bring it in line with the Paris Agreement.[12]

Attribution science may yet play a role in the outcome of international climate-related lawsuits, but enforcement of local environmental law is also heating up. Back in the UK, the government's own Environment Agency, for example, has called for more serious legal action to be taken against water companies for environmental crimes, calling in court to impose higher fines, to jail chief executives and board members whose companies are responsible for the most serious incidents and to strike off company directors so that 'they cannot simply move on in their careers after illegal environmental damage'.[13] The Department for Environment, Food and Rural Affairs and the Environment Agency announced the launch of the UK's largest ever investigation of all water companies into potential 'flow-to-full' treatment non-compliance, looking at whether they have knowingly or deliberately broken the law in relation to the treatment and discharge of sewage.

Food

One of the most direct interfaces between natural capital and the wider economy is food. This, in turn, is underpinned by energy, which, on the one hand, enables the human food chain and, on the other hand, can send shockwaves through it, and even dismantle it.

By May 2022, months after Russia's invasion of Ukraine, the world of agriculture was on the edge, even in the rich world. The UK's Andersons Centre tracks 'agflation' (rising food prices caused by increased demand for agricultural commodities) at 25.3 per cent in May 2022, compared to the already staggering 9 per cent CPI (consumer price index) inflation and 8.5 per cent CPI food inflation. In 2021, rising gas prices had pushed up the price of fertilizer. After Russia invaded Ukraine, sanctions hit supplies and so drove prices further. Diesel for farm vehicles had risen by 200 per cent and fertilizer costs by 400 per cent. The National Farmers' Union expected that 7 per cent of dairy farmers would stop producing milk. While producers were hurting, the pain was being passed to consumers, further fuelling the cost-of-living crisis.

The energy, carbon and wider ecological footprint of food sits behind these phenomena as a root cause. The food system is resource, energy and carbon intense, accounting for around 30 per cent of total energy consumption. It is run primarily on fossil fuels and thereby accounts for around 20 per cent of total global greenhouse gas emissions. Around 20 per cent of energy used in food production is associated with agriculture (around 60 per cent of which is from petrol, diesel, electricity, and gas and around 40 per cent from fertilizer and pesticide), around 14 per cent from transportation, around 16 per cent from processing and around 50 per cent from handling (such as from retail, restaurants, packaging, refrigeration, and preparation).

The most energy efficient foods include wheat, beans, fish, eggs and nuts. The least energy efficient foods are animal-based products such as beef, lamb, and goat. Beef, for example, requires up to 20 times more resources and emits 20 times more greenhouse gases than plant-based protein sources.

According to CarbonCloud, beef mince has a carbon footprint of 21 kg CO_2e/kg, compared to 19.9 for frozen prawns, 4.1 for a chicken breast and 1.8 for a plant-based burger. Plant-based alternatives to cow's milk such as from rice, oat and soya can cut emissions by over 50 per cent. (Paradoxically, rice is a bigger contributor to greenhouse gas emissions than any other food except beef and is responsible for as much greenhouse gas as aviation. Cultivating it deprives soils of oxygen and generates methane-emitting bacteria.[14]) Cheese has over three times the carbon footprint of chicken. Sadly, both chocolate and ice cream also have a higher carbon footprint than chicken, because of factors such as dairy and deforestation. In the sea, crustacean fisheries, such as those catching prawns and lobster, use the most fuel and generate the most emissions relative to their catch. But industrial fishing is creating other issues, now that around a third of the world's fish stocks are overexploited and in decline, while farms are polluting coastal areas with excess nutrients from fish faeces and uneaten feed, which can lead to oxygen depletion and destruction of local eco-systems.

Food supply is driven almost entirely by non-renewable sources and accounts for nearly 20 per cent of fossil fuel use in the USA. A study by Emory University estimated that it takes around 7.3 units of primarily fossil fuel energy to produce one unit of food energy in the USA. The journey starts in the production system. Oil (used world-wide in food, transport and heating) enables cultivation by running tractors, combine harvesters and other farm vehicles and equipment that plant, pump water, spray herbicides and pesticides, harvest and transport food and seed.[15] Food processors rely fresh or refrigerated food to be delivered to them by oil-based vehicles, as well as on the production and delivery to them of food additives, including emulsifiers, preservatives, and colouring agents, many of which are themselves oil-based. Then they rely on the production and delivery of boxes, metal cans, printed paper labels, plastic trays, cellophane, glass jars, plastic and metal lids and sealing compounds, many of which are essentially oil-based. Then they deliver finished food products to distribution centres, retail stores, restaurants, hospitals and schools, often in refrigerated vehicles. The 'last-mile' delivery of goods to homes and, vice versa, of consumer trips to shops for supplies adds the last gallons of oil to the food journey from the farm to the fork.

To give some sense of the magnitude of this, it's worth taking a moment to digest some of the implications of the journey so far. Absent a requirement for locally sourced food, a tomato can travel over 4,000 kilometres to end up in a convenience store aisle. Estimates expressed as volumes of oil equivalent per unit of food output are engaging. A Spanish tomato in a Scandinavian supermarket has an embedded production and transportation energy cost equivalent to around five tablespoons (each 14.8 ml) of diesel fuel for every medium-sized (125 grammes) tomato, equivalent to about one-sixth of a bottle of wine.[16] A dinner service of a kilogramme of reasonably efficiently produced roasted chicken (the dominant meat in all Western countries) has an energy cost equivalent to almost half a wine bottle of crude oil. A kilogramme of bread is around a third of a bottle. Somewhat inevitably, a new phrase is born: 'fossil food'.

Vertical farming and greenhouse solutions promise to increase the energy and resource efficiency of farming, cutting out many of these costs by producing food closer to the point of consumption and dramatically lowering levels of land take, resource and water use. Vertical farming seeks to grow fruits and vegetables in tightly controlled indoor environments. They seek to optimize plant growth and use soilless farming techniques such as hydroponics, aquaponics and aeroponics. You might find vertical farms in buildings, shipping containers, tunnels and abandoned mineshafts. The concept was the brainchild of a Public and Environmental Health professor at Columbia University, Dickson Despommier, who in 1999 proposed a skyscraper farm that could feed 50,000.[17]

LED technology is at the heart of the solution to replicate sunlight, while vertical farms eliminate many 'externality'-associated risks such as pests, flooding, and drought. They don't need pesticides or herbicides and use 90 per cent less water than traditional farming and recycle moisture for irrigation. Because they can be located almost anywhere, in principle in an industrial park by London or New York, they can cut down transport time and cost. There is, of course, no free lunch. Vertical farms need large amounts of energy, not just for the lights but for the heating and ventilation systems. Energy can represent around half of the operating costs. On-site generation and cogeneration offer potential solutions, particularly as a power failure could wipe out a crop, as do greenhouses.

But getting from the farm to the fork is not quite the end of the journey. What is not converted to energy for humans or animals, ending its journey as residue in the sewage system, is wasted. In the USA, 30–40 per cent of food is wasted. According to the Food and Agriculture Organization of the United Nations, the world loses almost half of all root crops, fruits, and vegetables, around a third of all fish, some 30 per cent of cereals and a fifth of all oil seeds, meat and dairy products. The UK's Waste and Resources Action Programme estimates that 70 per cent of wasted food was edible and was not consumed because either too much food was served or it spoiled. Spoiled food tends to end up in landfills and produces a large amount of methane – a potent greenhouse gas. Food waste is estimated to be responsible for 6–10 per cent of the world's greenhouse gas emissions. It is estimated that around one-quarter of the calories the world produces is thrown away, either spoiled or spilled in supply chains or wasted by retailers, restaurants, and consumers. A recent study found that almost a quarter, 24 per cent, of food-related emissions come from food that is lost in supply chains or wasted by consumers.[18]

Water

Food and food-related waste also feed back to the water supply. Agriculture and the food production process consume around 70 per cent of all water used. The food cycle is also one of the most significant causes of human-induced water pollution. Chemical dumping from agriculture causes eutrophication of water, killing fish and creating 'dead-zones'.[19] According to the United Nations, more than 80 per cent of the world's sewage (including food waste and faecal water) enters seas and rivers untreated. Fishing boats, tankers and cargo ships cause plastic and other pollution. Global warming heats water, reducing its oxygen content. Deforestation exhausts water resources and generates organic residues that breed harmful bacteria. Deteriorating water quality can stunt economic growth and exacerbate poverty in some countries. The World Bank has identified an edge, a threshold beyond which organic pollution in water reduces the growth in GDP of the affected regions by a third. More immediately, water pollution contaminates the food chain, which is harmful to health and reduces access to drinking water, a problem that now besets billions of people. The World Health Organization estimates that 2 billion people have no option but to drink water contaminated by excrement, exposing them to diseases such as cholera, hepatitis A and dysentery. According to the United Nations, diarrhoeal diseases linked to lack of hygiene cause the death of around 800 children a day worldwide.[20]

It is estimated that half of the world's population will live in water-scarce areas by 2025. Just as the solutions to energy and waste in the food system involve reducing – reducing miles through locally produced food, reducing packaging, reducing processing, reducing resource-intensive meat consumption, improving farming methods to reduce nitrogen and other resource use and reducing food waste sent to landfill – so prevention of water pollution involves reducing – reducing the use of chemical pesticides and fertilizers on crops, reducing and treating waste water so that it can be used for irrigation and energy production instead of polluting, reducing single-use plastics, reducing over-fishing and, over the longer term, reducing CO_2 emissions to prevent warming and acidification of the oceans. Resource efficiency and productivity gains that are good for both the environment and people, like in the energy sector, depend on reducing consumption for the same or better level of output. Like energy inefficiency, the inefficiencies associated with the food system and natural capital are, at once, a huge and far-reaching problem and a huge opportunity. Natural resources can be regenerative and renewable, but they can also be over-exploited, depleted and spoiled. There are limits.

Food and Fuel

There are other interfaces and limits between natural capital and energy, between food and fuel. After Russia invaded Ukraine, food prices soared, causing a cost-of-living crisis in the rich world and increased risk of famine in poorer countries. This raised pressure on producers of low-carbon fuels derived from crops, igniting a 'food versus fuel' debate that raged in the last major food crisis in 2007–8. Back then, the World Bank and the IMF attributed 20–50 per cent of the increase in the price of corn to competition from biofuels. The UN's food spokesperson had called this a 'crime against humanity'.[21] This time round, prior to the Russian invasion of Ukraine, global biofuel production was booming and had reached record highs.

The USA was not only the world's largest producer of oil and gas but also of biofuels. 36 per cent of its total corn production went to biofuels in 2021. Over 40 per cent of the soyabean oil produced in the USA in 2022 was used to make renewable diesel and other biofuels that year. But in 2022, biofuels once again came into competition with global grain and vegetable oil supplies. Russia and Ukraine produced nearly 20 per cent of the world's corn and more than half of its sunflower oil, but crop exports fell following the invasion, putting hundreds of millions of people at risk of 'hunger and destitution' from food shortages caused by the war, according to the United Nations secretary-general. To frame it, the total amount of crops used each year for biofuels is equivalent to the calorie consumption needs of 1.9 billion people.[22] The World Resources Institute estimates that a 50 per cent reduction in grain used for biofuels in Europe and the USA would compensate for all the lost exports of Ukrainian wheat, corn, barley and rye.

The environmental campaign group Transport & Environment estimates that the equivalent of 15 million loaves of bread is burned every day as ethanol in cars. However, there are two sides to every story. The Union for the Promotion of Oil and Protein Plants presents data to show that only a small share, some 2 per cent, of all cropland is allocated to biofuels, while the ethanol industry points out that most of the grain used to produce fuel is feed wheat, which goes into animal food, as opposed to milling wheat, which is made into bread. Meanwhile, the biofuel industry points out that it is itself a significant producer of animal feed through protein and fat by-products that are fed to chickens, cows and pigs. Sustainable land use might depend on what problem you are trying to solve, for who, where and when.

Footprints

At this point, it's useful to return to the concept of ecological footprint discussed towards the end of Chapter 2. The 2022 edition of the Global Footprint Network's National Footprint and Biocapacity Accounts, based on data from 2018, estimates that the world is in 'deficit' or 'overshoot' as it is consuming some 1.2 global hectares of biocapacity per person, some 70 per cent more than it produces.[23]

('Earth Overshoot Day' landed on 28 July in 2022. This day marks the date when humanity has used all the biological resources that the Earth regenerates during the entire year. This was late December in 1971.) This is the depletion of natural capital in action.

In summary, think of it like this, as Maurice Strong put it: 'After all, sustainability means running the global environment – Earth Inc. – like a corporation: with depreciation, amortization, and maintenance accounts. In other words, keeping the asset whole, rather than undermining your natural capital.'[24]

- Society depends on the ecosystem.
- The World Economic Forum characterized biodiversity loss in its 2022 Global Risks Report like this: 'Irreversible consequences for the environment, humankind, and economic activity, and a permanent destruction of natural capital.'
- Natural capital and human capital are inextricably linked.

CHAPTER 8

Human Capital

• • •

- Cities are key arenas for climate change and environmental degradation because most people live, and most resources are used, in cities.
- Cities face the brunt of extreme weather.
- Mitigating climate change raises major questions around income inequality.
- Resource intensity and extraction take a human toll.

When I started Sustainable Development Capital in 2007, it was the same year that Russia planted its titanium flag on the floor of the Arctic, laying its claim to a fifth of the world's natural gas reserves. It was the year that saw the beginnings of the Great Financial Crisis. It was also the year during which the world was to cross the historic milestone when, the UN estimates, for the first time, more people in the world lived in urban than in rural areas.

Cities are where we come together for food, shelter, communication, safety, culture and trade. They are the nest of human capital and the main destination of natural capital. Covering 2 per cent of the Earth's surface, they are responsible for some 75 per cent of resource consumption, waste and pollution and 80 per cent of GDP. They are the clearest manifestation of the choices of humanity. They personify civilization and society. They are where we will choose to fail or succeed.

By 2030, 85 per cent of the UK and US populations, 70 per cent of the Chinese population and some 50 per cent of Africa will live in cities. 1.5 million people move into cities around the world every week. By 2050, nearly as many people as are alive today, some 7 billion, are projected to live in cities. This will demand construction of around 2.5 trillion square feet of building space, double today's global building stock and equivalent to one New York City every 30 days for the next 40 years. Over 70 per cent of carbon emissions in New York come from buildings. Other than water, concrete is the most widely used substance on the planet and cement, its key ingredient, is the source of 8 per cent of the world's CO_2 emissions. It is estimated that China used more concrete in 2011–13 in the middle of the largest and fastest urbanization process in history than the USA did in the entire twentieth century.[1] Concrete and steel for urban construction generate the demand for most of the world's mining activity, while urban construction represents around 10 per cent of global CO_2 emissions. What goes up can also come down. Demolishing a single large building in a modern city can generate as much or more waste as a 10 million-person city produces in a year.

Buildings

Existing building stock will demand solutions such as district heating and cooling, demand-side flexibility and potentially heat pumps, that can reduce energy use by up to 50 per cent. Some economies and governments have responded to the demands from new building stock by boosting high-performance construction. Japan's Zero Energy Housing accounted for over 16 per cent of the private housing market in 2021, up from around 3 per cent in 2014. Likewise, China's commitment to green buildings construction has a target for more than 50 per cent of urban development to meet a green standard as part of the country's previous 13th Five-Year Plan. In 2021, energy efficiency investment in China was estimated at about US$35 billion, and total 'green' building floor space in China is estimated to be around 6.6 billion m^2, while total 'green' residential floorspace amounted to around 52 billion m^2.

Transport

Petrol cars, trucks, buses and other vehicles generate around 16 per cent of greenhouse gas emissions; 92 per cent of the world's population now lives in areas where the air quality is below the level considered safe by the World Health Organization. Air pollution associated with buildings (cooking and heating), industry, power generation and transport in cities are estimated to cause 8.7 million premature deaths a year. Air pollution in cities takes more lives each year than war, terrorism and murder combined. This, together with reducing reliance on oil for energy security reasons, is one of the key drivers of, and reasons for, government initiatives and incentives to transition to electric vehicles, sales of which doubled globally from 2020 to 2021. According to the UN IPCC, 'electric vehicles powered by low-emissions electricity offer the largest decarbonization potential for land-based transport, on a life cycle basis'. In the air and the sea, the UN IPCC states that sustainable biofuels, low-emissions hydrogen, and derivatives have the potential to reduce CO_2 emissions from transportation sectors but necessitate improved production processes and cost reductions. From what I can see on the ground, the change may happen sooner than many think.[2]

Sea Levels

Cities will be part of both the cause and effect of pollution and climate change. Over the longer term, sea-level rises threaten the existence of most low-lying coastal cities and many, such as Miami, have drawn comparison to Amsterdam in the need to prepare to be living below sea level, with huge sums invested in flood defences. The National Oceanic and Atmospheric Administration has warned that, by 2050, the heart of the USA's oil and gas industry and centre of global fossil fuel trade could itself be underwater half the year.

Heat

Cities, in many ways, are the front line. Cities are hotter because land cover has been replaced with pavements, buildings and other heat-absorbing surfaces, creating 'heat islands'. Heat kills around 350 New Yorkers a year and Black New Yorkers are twice as likely to die from heat as white residents. Economists have estimated that 'heat stress', reducing productivity during hotter months, could cost the USA alone US$200 billion a year by 2030.[3]

Right now, cities are bearing much of the brunt of the heat. In June 2022, Madrid endured temperatures of 40°C, and the UK matched it, hitting an all-time high of over 40°C. Vulnerable people died. Heatwaves in Spain now kill between 1,500 and 1,700 people every year. At the same time, record daily temperatures were being registered in China and the USA. In Nevada and Arizona, Lake Mead's water levels sank to an all-time low, putting drinking supplies for 25 million people at risk. The rise in deaths from heat is projected to outweigh any falls in deaths from cold and affects poor communities, who cannot afford cooling and who suffer from urban 'heat island' effects more severely. Extended heatwaves threaten to reduce the supply of electricity, for instance from hydro, and to increase demand for cooling, straining the electrical grid. This can disrupt the food supply chain and leave buildings and residents stranded from power outages.

The impact will be most severe in hotter and poorer countries. Most of the global population lives in tropics and sub-tropics. Scientists estimate that the maximum heat at which people can survive is 35°C. If they can't survive, they will move or die. We have seen mass migration from conflict following the Arab Spring, from drought and competition for resources from Darfur, Syria and Mali. Currently, climate-induced migration is relatively low as fewer than 1 million people live in areas that average between 38°C and 45°C in the shade during the hottest months. But even if we are successful in limiting global temperature rise to 1.5°C, that number will rise to 30 to 60 million this century, according to the International Organization for Migration. In scenarios above

1.5°C, the numbers become several hundred million. The epicentres will include West Africa, the Middle East, and South Asia, where India and Pakistan hit new records in 2022, and suffered deaths from severe weather events with few natural capital defences. Tragically, by August 2022, 10–12 per cent of Pakistan was under water and 33 million had been affected by historic levels of flooding, which would kill nearly 2,000 people and cost some US$30 billion or 10 per cent of GDP. The northern region of Pakistan, sometimes referred to as the 'third pole', contains the world's largest amount of glacial ice outside of the two polar regions. Glacial melting in the Gilgit-Baltistan and Khyber Pakhtunkhwa regions has, according to the UN Development Programme, created more than 3,000 lakes, 33 of which are at risk of sudden bursting, which would put 7 million people at risk. By September 2022, Somalia was facing famine.

Meanwhile, cold and fuel poverty kill too. In the UK, an average of 9,700 deaths a year have been attributed to living in a cold house. Over 5 million deaths a year have been attributed to excessive heat or cold.

Income Inequality

The impact of environmental degradation and climate change will be felt hardest by the poor. This is not just an argument between rich and poor countries, between developed and advancing economies. Climate justice and equality is a matter of furious debate at climate conferences, despite spending promises from the rich world to the poorer countries of US$100 billion a year to address it. However, the richer countries of the West and the North face domestic emerging markets of their own. Income inequality has spiralled in the last 40 years. According to the World Business Council for Sustainable Development, nearly 30 per cent of all growth from 1980 to 2016 was captured by only 1 per cent of the global population. Around 12 per cent of growth was captured by the 'bottom 50 per cent', in the rise of the emerging economies. Real incomes, on the other hand, fell for the 'squeezed bottom 90 per cent' in the USA and Western

Europe. All told, in rich and poor countries, the World Bank estimates that climate change will push up to 130 million people into poverty in the next 10 years.

Climate justice is indeed one of the most contentious and explosive issues at stake. Climate change mitigation policies are held out as tools for poverty alleviation and, at the same time, translate into demands for the poorest populations to take actions that they can afford the least. This infuriates developing economies, as well as poor communities in the USA and the UK, which argue that climate policies can make life worse for the poor. Poor societies start off being more vulnerable to natural disasters. Money invested in bringing poor people out of poverty and sheltered and housed in buildings that withstand severe weather would, they argue, generate substantially larger and faster benefits to them than longer-term climate policies aimed at reducing CO_2 to the same end. Poorer communities are less able to invest in adaptation to protect lives and goods.

Comparison is made between the wealthy Netherlands and the poor Bangladesh. The Netherlands has been able to respond to floods because it has access to capital. Responding to events such as the 1953 *Watersnoodramp* that flooded 9 per cent of farmland, drowned 30,000 animals, destroyed 10,000 buildings and killed 2,551 people, the Netherlands invested US$11 billion in the 'Delta Works', a system of dams and surge barriers. Since then, there have been only three floods and one fatality. Bangladesh, with far less financial resource, deforesting while the Netherlands plants trees, loses thousands of lives to flooding and millions more are displaced. Meanwhile, more people, it is argued, could be pushed to starvation by higher energy prices exacerbated by climate policies than would be pushed to poverty from lower food production and higher food prices caused by climate change. Food price hikes in 2008 were at least partially attributed to the growth in biofuel at the time, in what the UN special envoy for the right to food, Olivier De Schutter, described as a 'silent tsunami' that had pushed 100 million into poverty and 30 million into hunger. It is estimated that the same US$100 billion a year that is currently pledged for climate investment could lift everyone on the planet outside of extreme poverty.[4]

When the rich world asks the poor world to pursue a renewable energy pathway, it can sound like a demand to take a more expensive pathway with less energy supply or security, to sacrifice poverty alleviation on the altar of more wealth creation (or waste) for the rich. As it stands, electricity provision to Africa is currently falling because of the economic impact of the COVID-19 pandemic and Russia's invasion of Ukraine. The International Energy Agency estimates that annual investment of US$25 billion, or 1 per cent of total global energy investment, would reverse this and deliver universal energy access to Africa by the end of the decade. Africa receives around 7 per cent of the total 'climate finance' that flows from rich countries to poorer ones. Out of 600 million people, 43 per cent lack access to electricity. Renewables can play a big part in this; the IEA estimates up to 80 per cent. Africa has around 60 per cent of the best solar resources on Earth but has only 1 per cent of installed solar capacity. But it would still also need to produce an additional 90 billion cubic metres a year of natural gas by 2030 to industrialize, for example for producers of fertilizer, steel and cement, as well as water desalination. Although an IEA report in 2021 called for no new fossil fuel developments in the pathway to net zero by 2050, Fatih Birol, Executive Director, stated in 2022 that it would be fair for Africa to develop hydrocarbons for industrial use: 'You cannot use wind or solar, at least now, to build those industries.' Natural gas, meanwhile, generates only around half the greenhouse gas emissions as coal.[5]

The UN IPCC concludes that climate change disproportionately affects vulnerable people and ecosystems due to socio-economic factors and unsustainable practices.[6] Further, that urbanization offers a critical opportunity for climate-resilient development through inclusive planning and investment in infrastructure, benefiting marginalized communities and ecosystems.[7]

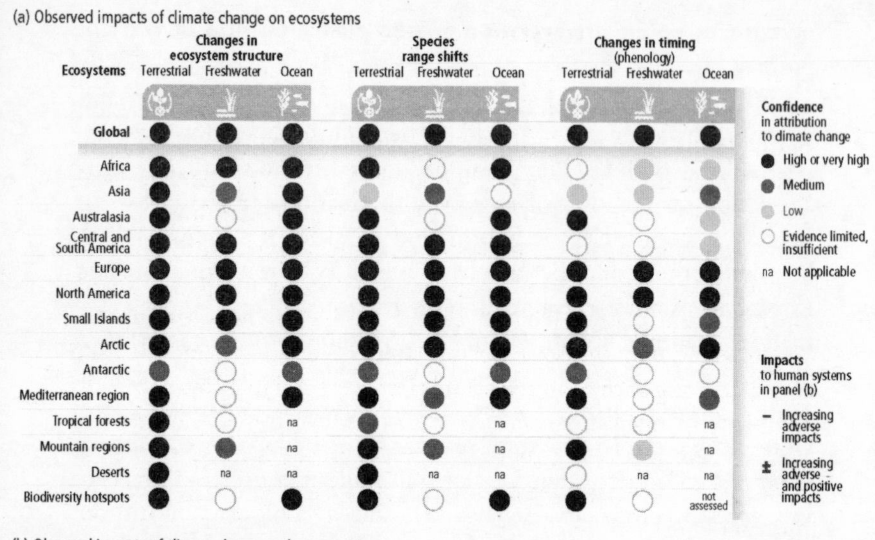

(b) Observed impacts of climate change on human systems

Impacts on water scarcity and food production: Water scarcity; Agriculture/crop production; Animal and livestock health and productivity; Fisheries yields and aquaculture production

Impacts on health and wellbeing: Infectious diseases; Heat, malnutrition and other; Mental health; Displacement

Impacts on cities, settlements and infrastructure: Inland flooding and associated damages; Flood/storm induced damages in coastal areas; Damages to infrastructure; Damages to key economic sectors

Human systems: Global; Africa; Asia; Australasia; Central and South America; Europe; North America; Small Islands; Arctic; Cities by the sea; Mediterranean region; Mountain regions

not assessed; na

Figure 8.1 Impacts of climate change are observed in many ecosystems and human systems worldwide

Source: Figure SPM.2 from IPCC, 2022: Summary for Policymakers [H.-O. Pörtner, D.C. Roberts, E.S. Poloczanska, K. Mintenbeck, M. Tignor, A. Alegría, M. Craig, S. Langsdorf, S. Löschke, V. Möller, A. Okem (eds.)]. In: Climate Change 2022: Impacts, Adaptation, and Vulnerability. Contribution of Working Group II to the Sixth Assessment Report of the Intergovernmental Panel on Climate Change [H.-O. Pörtner, D.C. Roberts, M. Tignor, E.S. Poloczanska, K. Mintenbeck, A. Alegría, M. Craig, S. Langsdorf, S. Löschke, V. Möller, A. Okem, B. Rama (eds.)]. Cambridge University Press, Cambridge, UK and New York, NY, USA, pp. 3–33, doi:10.1017/9781009325844.001. https://www.ipcc.ch/report/ar6/wg3/downloads/report/IPCC_AR6_WGIII_FullReport.pdf

Resource Intensity

As ever, energy, particularly renewable energy, grabs the sustainability headlines. But in some respects, renewable energy can be seen as being part of a solution to an underlying problem – that of large, increasing and carbon-emitting resource use caused by human consumption patterns. The consumption associated with human capital is driven by the use of key resources used by modern society, including steel, cement, chemicals and plastics (collectively responsible for more greenhouse gas emissions than the power sector), and key inputs into industry and manufacturing such as ammonia (for fertilizer) and nickel, cobalt, lithium, copper, manganese and rare earths (part of the basis of modern industry and the low-carbon economy). And the extraction of these resources has other far-reaching implications for human capital, and human rights.

Take one example, cobalt, used in electric car batteries. The Democratic Republic of Congo's 'Copperbelt' provides some 60–70 per cent of the world's cobalt, most of which is a by-product of large copper mines. But the rest of it comes from small-scale informal 'artisanal' miners who are responsible for up to 15 per cent of Congo's cobalt.[8] To give some sense of scale, that is more than the entire output of Russia, which is the second largest producer in the world (Australia being the third). Artisanal miners hand dig high-grade ores. But small mines are dangerous and polluting (fathers that mine there are more likely to have children with birth defects). Miners are killed when tunnels collapse, or fires break out underground. NGOs have reported that children as young as seven were found digging in the mines. According to the OECD and BGR, children work in 25–30 per cent of mines. Up to 200,000 people work as artisanal miners in the Copperbelt, and perhaps up to 60 per cent of households in the region depend on it. The energy transition will need to bring people along with it.

Meanwhile, at the other end of the supply chain, a combination of simultaneous resource intensity, resource scarcity, supply chain disruption, price increases, rising human and transport costs and geopolitical conflict has fuelled global inflation. The prospective impact of permanently higher prices on human capital will play out over the coming years but will be affected by a new era of pressure on costs and security of supply, with impacts on the cost of living and, for some, poverty. Efficiency will be key.

The Consumer

Human capital is the ultimate decision maker as it is the source of all demand and political licence. Consumer engagement with environmental issues delivers a demand signal to the market and the supply chain that encourages or forces them to meet the needs. If the consumer wants sustainably sourced food, fabrics, or fabrications, then the market will deliver them. Any price differentials will be spread between the supplier and the consumer. Governments may help bridge the gap in cost or set legal, regulatory, or labelling standards.

This 'demand pull' from human capital takes time but can have a major impact. Changes in consumer demand patterns have helped displace straws, carrier bags, and other single-use plastics from mainstream use in the UK. The Top Runner programme for energy efficient appliances in Japan is a manifestation of a cultural consumer preference for sustainable products. Some of the most successful business stories in the food industry have resulted from satisfying the growing consumer appetite for eco-friendly meat alternatives. The global fashion industry has seen the rise of ethical or eco-friendly brands, as well as a trend towards 'circularity', that is the repair, redesign, rental, swapping, re-selling, or recycling of garments, most of which would otherwise be disposed of and wasted.

A sector where there is likely to be significant consumer demand for change is automotive. Consumers are buying electric cars as part of a global community fighting climate change and pollution. The associated extraction of scarce resources and rare earths, together with the use of more materials than business as usual, a potentially high carbon footprint and labour issues all conflict with much of the original intent to actively participate in healing the planet, not to make it worse. Consumer awareness drives change and the car manufacturers selling to them know that, hence their increased focus on sourcing minerals and metals more sustainably. It is therefore in the manufacturers' best interests to understand the environmental and social impact of every material in their supply chains. BMW, Mercedes-Benz, Ford, General Motors, Volkswagen, and Tesla have all now signed on to the Initiative for Responsible Mining Assurance (IRMA) to understand their sourcing, and the risks.

Demands from the market and civil society can often work faster and more universally than regulatory levers.

Calls to Action

Similarly, cities have demonstrated an ability and willingness to act, in many cases, faster than national governments. Over 700 cities in more than 50 countries have committed to halve greenhouse gas emissions by 2050 in the 'Race to Zero' initiative. More than 50 cities have published Climate Action Plans with objectives to take action that would be aligned with a 1.5°C implied temperature rise. The C40 Cities network estimates that these actions could avert more than 1.9 gigatonnes of emissions by 2030, equivalent to half of the annual emissions of the European Union.

While consumers will establish the needs that the market will fulfil, the actors, particularly for change, are a smaller number of people and leaders directing the metaphorical traffic, in civil society, the media, business, government and finance. This is human capital. And as the anthropologist Margaret Mead once said: 'Never doubt that a small group of thoughtful, committed citizens can change the world. Indeed, it is the only thing that ever has.'[9]

- Cities are home to 75 per cent of resource consumption, waste and pollution and 80 per cent of GDP.
- Climate change mitigation strategies confront issues such as income inequality.
- The human impact of supply chains and the need for resource efficiency are becoming better understood.

CHAPTER 9

Sustainable Finance, ESG, Net Zero and Greenwash

• • •

- The extraordinary momentum behind ESG up to 2021 met with challenges and controversy in 2022 and, to some degree, went into reverse.
- The Russia–Ukraine crisis made ESG harder.
- Sustainable finance became open to abuse.
- Better approaches to and measurement of positive financial, economic and social impact should emerge.

Just as things are hotting up, the world of 'sustainable' finance bumps up against the edge.

By 2021, 'ESG' assets, that is funds under institutional management associated with environmental, social or governance objectives, were growing at 30 per cent a year, had exceeded US$37 trillion and, according to Bloomberg, were on track to exceed US$50 trillion or one-third of all assets under management by 2025. By 2022, this was heading into reverse. (In addition, there had been some high-profile instances of financial investors using their firepower to effect change in the old as well as the new. In May 2021, Exxon shareholders voted to fire two board members after an activist investor, Engine No. 1, complained that they weren't taking climate change seriously.)

At stunning speed, the limits seem to have been reached and boundaries crossed. May 2022 saw the largest ever outflows from

ESG funds. The spike in oil and gas prices since Russia invaded Ukraine lifted fossil-fuel shares, driving the S&P 500 Energy Index to gain 59 per cent, at the same time as the benchmark overall dropped 14 per cent. This challenged the concept that ESG funds could deliver the same returns or better than traditional benchmarks. As the effects of the Ukraine crisis were felt in Europe and North America, the narrative was changing. Divesting of fossil fuel companies from institutional portfolios was no longer vogue. The narrative changed seamlessly and unnervingly elegantly. Put simply, we won't sell fossil fuel companies because we care about the environment, and we don't want to sell to people who don't. Malaysia's sovereign wealth fund said: 'It's easier for some multinational funds to say: I will sell your company and I will abandon you … [but] … We are not there to pull the rug from under their feet.' LNG in the USA, one of the most important alternatives to Russian gas for European energy security, was suddenly branded as 'the largest green initiative on the planet, equal to the combined impact of every domestic mainstream green solution'.[1] According to a marketing document produced by natural gas producer EQT, by 2030, an unleashed US LNG scenario would reduce international CO_2 emissions by an incremental 1.1 billion metric tonnes per year, equivalent to electrifying every US passenger vehicle, powering every home in America with rooftop solar and backup battery packs, and adding 54,000 industrial-scale windmills, doubling US wind capacity.

On 30 June 2022, the US Supreme Court ruled that the Environmental Protection Agency did not have authority to limit emission across whole states. The case was brought by West Virginia on behalf of 18 other mostly Republican-led states and some of the largest coal companies in the USA. The complaint was that emissions reduction policies would cause economic and job losses. The 19 states made up 44 per cent of the USA's emissions in 2018 and had achieved on average only a 7 per cent reduction in emissions since 2007.

In July 2022, Rishi Sunak felt that it would be an election winner for him to pledge to ban onshore wind if he were elected UK Prime Minister (albeit offering to double-down on a longer-term offshore wind programme).

Lobbies

As the wind direction changed, lobbies kicked in and legal and political narratives abounded. In June 2022, the Business Roundtable, a lobby group representing CEOs of some of the largest companies in the USA, pivoted. In 2019, most of the group's members had pledged to 'protect the environment by embracing sustainable practices across our businesses'.[2] Now it was pressing the SEC, the financial regulator, to halt its climate rule that would force companies, including real estate (responsible for 40 per cent of greenhouse gas emissions), to disclose carbon emissions and climate change risks for the first time, branding it as 'unworkable'. The US Chamber of Commerce joined the chorus and called for the rule to be weakened. The estimated combined spend of both lobby groups in Washington is estimated to be approximately US$50 million a year, so powerful voices.

States

On the legal front, the Kentucky Attorney-General released an opinion on 26 May 2022 that investment managers who are responsible for pension funds and who simultaneously pursue ESG strategies are not acting with fiduciary responsibility and are instead acting with political agendas and partisan beliefs: 'Today, in perhaps an even more pernicious version of the trend, the debate is no longer left to stockholders. In fact, there is little-to-no debate. Investment managers in some corporate suites now use the assets they manage – that is, other people's money – to enforce their preferred partisan sensibilities and to seek their desired societal and political changes.' ESG, associated in the opinion with an intention to pursue a social purpose or 'sacrifice some performance on their investments to achieve an ESG goal', was declared 'inconsistent with Kentucky law governing fiduciary duties'. In Idaho, legislation went into effect in July 2022 prohibiting the state's investment funds from considering ESG characteristics 'in a manner that could override the prudent investor rule'.[3]

In 2021, a law had been passed in Texas prohibiting state agencies from investing in businesses that had cut ties with fossil fuel companies. West Virginia passed a similar law in March 2022. By the summer, Kentucky, Tennessee and Oklahoma passed similar laws and New York, Oregon and Virginia were working on them. On 23 August, Governor Ron DeSantis and the Trustees of the State Board of Administration passed a resolution directing Florida's fund managers to prioritize return on investment without consideration for the 'furtherance of social, political, or ideological interests', or 'whimsical notions of a utopian tomorrow'.[4] By the following May, DeSantis had signed a bill requiring the Florida State Board of Administration, Tallahasee, to make investment decisions based 'solely on pecuniary factors' – or in other words, not on ESG or 'ideological' factors.

Republican state treasurers launched coordinated efforts to thwart and punish climate action at both state and federal levels. In the summer of 2022, Riley Moore, the treasurer of West Virginia, announced that several major banks, including Goldman Sachs, JP Morgan and Wells Fargo, would be barred from government contracts on the grounds that they were reducing their investments in coal. Together with the treasurers of Louisiana and Arkansas, they pulled more than US$700 million of investment funds under management out of BlackRock, complaining that it was too focused on environmental issues. The treasurers of Utah and Idaho put pressure on the private sector to drop climate action and other apparently 'woke' causes. The treasurers of Pennsylvania, Arizona and Oklahoma joined a campaign to thwart the nomination of federal regulators who wanted to require disclosure of financial risks from climate change by financial institutions. *The New York Times* identified the State Financial Officers Foundation at the nexus of the operation.[5]

Central Government

By June, Republicans in Washington had seized on ESG investing as a political issue. Republican senator Steve Daines from Montana indirectly joined the assault on the SEC's climate rule by questioning whether HSBC (my former employer) faced pressure to suspend the head of responsible investing for its asset management division after an apparently pre-approved presentation at a *Financial Times* conference at which he told investors not to worry about climate risk and that 'there's always some nut job telling me about the end of the world'.[6] Senator Daines wrote to the HSBC chief executive to say that he was 'concerned that this episode may involve breaches of US law [if the] ... suspension was in response to pressure on HSBC from outside parties that may be legally prohibited from influencing the management of your company'.[7] The letter even went on to postulate who some of the 'outside parties' might be as some of the largest institutional investors in the world.[8]

Investment Management

Little in the way of a counterattack was being mounted by the investment management industry. It was in defence mode and taking shots from the regulator. Only months before, investment management firms had publicly committed to coordinating joint action for ESG purposes, such as reducing climate change. For example, the Steering Committee for the Glasgow Financial Alliance for Net Zero (GFANZ) stated: 'The systemic change needed to alter the planet's climate trajectory can only happen if the entire financial system makes ambitious commitments and operationalizes those commitments with near-term action. That is why we formed [GFANZ], to bring together over 450 leading financial enterprises united by a commitment to accelerate the decarbonization of the global economy.'[9]

Similarly, Climate Action 100 'aims to ensure the world's largest corporate greenhouse gas emitters take necessary action on climate change',[10] stating that its investor signatories believe that taking action 'is consistent with their fiduciary duty and essential to achieve the goals of the Paris Agreement'. By June 2022, the US Securities and Exchange Commission (SEC) was investigating Goldman Sachs' asset management over ESG claims made by its funds. In May, BNY Mellon had become the first asset manager to settle with the agency for allegedly misleading investors about ESG claims. And Senator Daines was asking HSBC's CEO whether he had spoken to members of climate-focused industry groups, such as GFANZ or the Net-Zero Banking Alliance, over its suspension decision.[11]

In June, the head of Germany's largest asset manager, DWS, resigned after a police raid over claims that it has misled investors on its environmental record. Even within the sustainable finance industry, accusations of 'greenwashing' were flying. The Net Zero Asset Managers initiative, signed up to by 273 managers overseeing some US$60 trillion of assets, had established an agreement to set interim targets for emissions reductions by 2030. Sasja Beslik, chief investment officer at NextGen ESG Japan, a sustainable investment specialist, says: 'The NZAM commitments are purely aspirational and lack any detail explaining how the objective of decarbonising their investment portfolios will be achieved.' He dismisses the NZAM initiative as 'just a beauty parade'.[12]

By early 2023, major financial institutions had become pitted against their investment management shareholders. In April, some 30 per cent of shareholders backed resolutions, against board resolutions, for climate transition plans at Goldman Sachs, Wells Fargo and Bank of America.

Contagion had spread to industry. In July 2022, ClientEarth and Reclaim.NL sued Dutch airline KLM, claiming that its advertisements promoting the company's sustainability initiatives were misleading and constituted 'greenwashing'. The complaint claimed that KLM's carbon offsetting scheme creates the false impression that its flights won't make climate change worse.[13] KLM denied the allegations, arguing that its climate goals were credible – backed up investment of millions of euros in a more sustainable fleet and working towards the industry goal of net zero emissions by 2050 – and that it therefore had a right to advertise them. KLM launched an advertising campaign in December 2022, claiming to help pioneer a 'sustainable future' for aviation by paying towards reforestation projects or by contributing to the cost of greener aviation fuels, through its CO2ZERO programme. But campaign groups allege that aviation cannot be made sustainable quickly enough to meet global climate goals through replacing fossil jet fuel, improved engines, efficiencies, or other future technologies and want a reduction in flights. The case was heard in April 2023 and KLM took the opportunity to tell the court that it had pulled its 'Fly Responsibly' campaign, but it made no assurances about future advertising campaigns and maintained the carbon offsetting programme.

Greenwashing accusations made their way all the way up to government. In July 2022, the UK government was ruled by the High Court to have broken its legal obligations under the Climate Change Act because the government's net zero plan provided insufficient detail as to how the target would be met. The government was ordered to publish new plans that did by March 2023. In 2021, in a case brought by Greta Thunberg's movement, Fridays for Future, Germany's Federal Constitutional Court ordered the German government to improve its climate action plan. It said that it was insufficient to protect future generations and would have to be amended by the end of 2022.

By October 2022, the UK's financial services regulator, the Financial Conduct Authority, had to step in with proposals to introduce new rules around sustainable investment labels and disclosure.

What went wrong?

The Backstory

The 20 or 30 years in which 'ethical' investment was associated with underperformance were followed by a 'cleantech' boom and bust between 2003 and 2008. This had little to do with 'green hydrogen', batteries, heat pumps or much of today's technological zeitgeist, because this took place before the last decade and, by and large, involved backing of many of the right technologies too early and in the wrong places. The financial crisis of 2008 erased much of the gains from solar and electric vehicles of the previous five years, but it was fluctuating commodity prices, cheap natural gas and Chinese competition for solar panel production undercutting Western production that burst the bubble.

The 2009 COP in Copenhagen was a policy low point. Over the next decade, driven by a combination of tumbling prices for renewables, market incentives and government subsidies, and the scaling up of new solutions such as offshore wind, the market for clean energy infrastructure took off in earnest, with investment levels soon surpassing US$500 billion a year and attracting more new capital than conventional energy generation. Nearly a decade after Al Gore's seminal *An Inconvenient Truth*, the Paris Agreement represented a policy high point. The world seemed to galvanize around a target to limit global temperature rise to 1.5°C and it seemed to have the backing of government, business and society alike. This inspired years of growth for ESG and climate investment, led by ever cheaper, larger-scale and more cost-competitive renewable energy power plants visible from the highway and personified, on the highway itself, by the introduction of electric vehicles for the mass market.

Great Expectations

Green capital markets boomed, as did carbon markets. Green bond issuance broke into the trillions, where proceeds were earmarked for green projects of various shades. In all, the Climate Bond Initiative (for which I may be, ashamedly, the least active Advisory Panel member) has counted US$1.6 trillion of issuance since 2014 (peaking at 12 per cent of all bond issuance at the end of 2021), 80 per cent of which was designated for green projects in energy, buildings, and transport. As fear gripped markets, so did greed and, with it, perhaps some bad behaviour. Raising capital for purportedly green projects had, in certain areas, become just too easy and maybe a little lazy and cynical. The DWS ESG reporting and integration system, stated its whistleblower, was simply mis-stating the facts based on a mis-calibrated model-based system, partly because there was no measurement or tracking system, so no evidence.

Greenwashing

Some insight can be gained from studies conducted based on anonymized surveys of investor opinion. One study surveyed 1,130 investors in 19 countries and regions to get a sense of investor attitudes.[14] A quarter of investors said that ESG was central to their investment approach, but most described their approach as one of 'acceptance' or 'compliance'. Investment groups were giving clients what they wanted. When the survey asked whether investors' approach to ESG was driven by the expectations of client and reputational concerns rather than deeply held beliefs, the largest proportion of respondents agreed, 42 per cent, compared to those that disagreed, 23 per cent. When asked whether asset managers predominantly use ESG as a marketing and PR tool to generate sales and boost their reputation, 55 per cent agreed and 14 per cent disagreed. The biggest reason for adopting ESG was 'a desire to meet client needs and requests', at 27 per cent, a little more than the 25 per cent that said that it was a 'desire to do good'. 10 per cent said it was to help manage risks. Around half of respondents believed that 'greenwashing' was prevalent. 52 per cent agreed that greenwashing could trigger the next mis-selling scandal, versus only 14 per cent who disagreed. The biggest fix? Harmonizing global standards.

Standards

One problem is that there may just be too many 'standard' analytical frameworks for any of them to be genuinely useful. A study produced by the CFA Institute into ESG ratings of 400 companies in 24 sectors provided by MSCI, S&P, Sustainalytics, Carbon Disclosure Project, Institutional Shareholder Services and Bloomberg found the results to be 'conflicting and contradictory'. For example, MSCI's correlation with S&P was found to be below 50 per cent. By comparison, the same study looked at long-term debt ratings for the same companies and found that S&P, Moody's and Fitch ratings showed correlations of between 94 per cent and 96 per cent. At COP26, when MSCI presented its benchmarking methodology to rate companies based on the 'implied temperature rise' that their stated climate policies aligned them to, ESG specialists from institutional investors were intrigued and, in the meetings I attended, praised the work but asked them politely not to hold it out quite yet as a finished product.[15]

Indeed, some rating methodologies have produced some, at best puzzling, results. British American Tobacco was rated as the third highest ranked ESG company in the FTSE 100 Index in 2021 by Refinitiv, a subsidiary of the London Stock Exchange Group. In 2022, US electric car maker Tesla was ejected from the S&P 500 ESG Index. At the same time, S&P Global Ratings published ESG credit indicators as part of its credit ratings for states and state subdivisions. Russian energy giants Gazprom and Rosneft scored higher than Exxon Mobil. The Dow Jones Sustainability World Index, according to S&P Global, which produces it, 'comprises global sustainability leaders … the top 10 per cent of the largest 2,500 companies in the S&P Global BMI based on long-term economic, environmental and social criteria'.[16] As of 30 June 2022, none of the top 10 constituents was in the energy business, nor any directly related climate business. (The constituents were Microsoft, Alphabet, UnitedHealth Group, Taiwan Semiconductor Manufacturing, AbbVie, Roche, AstraZeneca, Novartis, ASML and Abbott Laboratories.) The MSCI ACWI ESG Leaders Index 'provides exposure to companies with high Environmental Social and Governance (ESG) performance relative to their sector peers'. As of 30 June 2022, only Tesla was in the energy or climate business, noting that it had been booted out of the S&P 500 ESG Index. (The constituents were Microsoft, Alphabet (twice), Tesla, Johnson & Johnson, Taiwan Semiconductor Manufacturing, Nvidia, Procter & Gamble, Visa and Home Depot.) Is this what investors think they are buying into?

Then there are standards that are aimed at helping companies set decarbonization goals and key performance indicators. Three 'leading' standards are the Task Force on Climate-related Financial Disclosures' (TCFD) Guidance on Metrics, Targets and Transition Plans, the Science-Based Targets initiative's (SBTi) Corporate Net-Zero Standard, and CDP's recently released Climate Disclosure Framework for Small and Medium-Sized Enterprises (SMEs).

Then there is regulation. For example, the Sustainable Finance Disclosure Regulation (SFDR) was introduced by the European Commission to 'improve transparency in the market for sustainable investment products, to prevent greenwashing and to increase the transparency around sustainability claims made by financial market participants'.[17] The SFDR, a cousin of the Corporate Sustainability Reporting Directive (CSRD), is part of the EU Sustainable Finance agenda, itself introduced as part of its 2018 Sustainable Finance Action Plan, which also includes the Taxonomy Regulation (a classification regime, which, amongst other things, declared gas and nuclear green in July 2022) and the Low Carbon Benchmarks Regulation. On the global scene, the Financial Stability Board, endorsed by the G20, created the Task Force on Climate-related Financial Disclosure (TCFD), which sets standards for climate-related disclosure and financial reporting. The G7 committed to mandatory disclosure and to support the International Financial Reporting Standards (IFRS) to establish a new International Sustainability Standards Board (ISSB) which published its framework in June 2023. However, in May 2022, the landscape had already become sufficiently complex that the Securities and Exchange Commission, the ISSB and TCFD set up a new workstream for 'convergence'.

If you are in business or the finance industry, study the 17 UN Sustainable Development Goals, decide which ones you align with and disclose it. (Note that none of today's 17 UN Sustainable Development Goals explicitly referas to resource efficiency.) If you didn't have a full-time professional ESG department, or an external specialist consultant, you do now.

The standards, rules, regulations and ratings make companies think. They have to think about their organization's governance around climate-related risks and opportunities. They have to disclose the 'actual' and 'potential' impacts of climate-related risks and opportunities on the organization's businesses, strategy and financial planning. They have to disclose how the organization identifies, assesses and manages climate risk. And they have to disclose the metrics and targets they use. However, standards and ratings methodologies do not, of course, determine what the companies do, or indeed do about ESG in general or climate change in particular. They are, for the most part, accounting disclosure related. They don't and can't ask companies to do anything to actively improve the environment. As the *Harvard Business Review* and Bloomberg recently put it, ESG ratings 'don't measure a company's impact on the Earth and Society. In fact, they gauge the opposite: the potential impact of the world on the

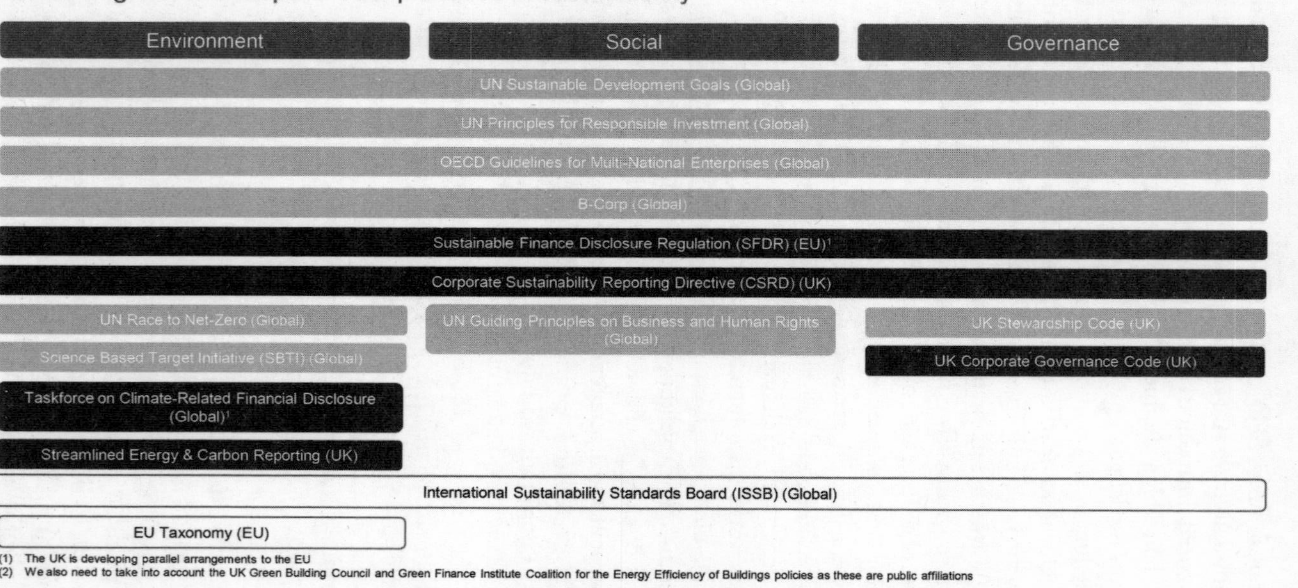

Figure 9.1 Bottom up: the sustainability ecosystem

Source: Sustainable Development Capital LLP (2022)

company and its shareholders'.[18] However, can they not ask different questions? Can they not require disclosure of how much energy and other resources are being wasted? Would shareholders not be shocked to find that their investments are tied up in companies losing a fortune in wasted energy, which in turn is producing needless excess carbon emissions? Why don't they measure and benchmark corporate energy efficiency? Standards, disclosures and ratings are a good start, but it is important that the theory survives contact with practice.

To illustrate the point, look at the European green taxonomy, which is meant to guide capital towards environmentally sustainable uses. It applies to the whole economy, regardless of the role a company plays in the energy or resource value chain. So only 2 per cent of the revenues of Europe's top 50 companies would be judged to have come from green operations.[19] For banks, loans to small- and medium-sized business or non-EU clients are excluded (so up to half of the assets of many large European banks). Meanwhile, investments or activities are either green or not, it is binary. It has been pointed out that a financing to upgrade a nineteenth-century building from the worst to the second-best energy efficiency category would not count as green, notwithstanding the fact that this would have a larger impact on emissions than financing a new build.[20] There is also a danger that it promotes 'paper decarbonization', whereby investors avoid high emitters to get a green label rather than engage in real-world efforts to reduce carbon.

Aiming Low: Net Zero

Meanwhile, climate objectives, when they are set out, tend to coalesce around either a net zero carbon target to be achieved by, say, 2050, or else a set of policies which if implemented would align with an implied temperature rise of no more than 1.5°C per the Paris Agreement and COP26. The UK government, for example, now requires listed companies and large asset managers to publish a plan for achieving net zero carbon emissions or else explain why not. While businesses grapple with their targets and what they are going to do about them, they also try to put further definition on what net zero itself means. Given that it is not possible to operate without greenhouse gas emissions, the onus must come down to reducing and removing. If there is any room for confusion in the corporate world, then this is reflected in the wider community and population. A poll of 2,000 people conducted in 2022 suggested that 57 per cent of people in the UK did not know what net zero is.[21] It is intriguing to consider this at the same time as the fact that net zero by 2050 is enshrined in UK law. The solution for listed companies and asset managers? Another watchdog, the Transition Plan Taskforce.

The problem is not limited to individuals and business. When top UK council officials were asked how confident they were that they would meet their own net zero targets, only 47 per cent said that they were on track.[22] To put this into context, 64 per cent of UK councils have committed to net zero by 2030. (One prominent UK mayor told me over breakfast in June 2022 that he refused to subscribe to a 2030 target, notwithstanding his deep convictions on climate, because he felt that it was simply unrealistic.) And as for central government? In June 2022, the UK government's official climate advisory watchdog presented its report to Parliament. Of 50 key areas for action, only 8 were deemed to be on track to deliver targeted levels of emission reduction. As it put it bluntly in its headline: 'Current programmes will not deliver Net Zero.'[23] While it found that the government had done quite well in reducing emissions from electricity generation, largely through deployment of renewables, and noted that electric vehicle adoption was on the rise, it pointed to a 'shocking gap' in policy on energy efficiency, while the 'weakest policies' applied to agriculture and land use. Perhaps levels of understanding as to what net zero actually means and how to get there are an issue at the top levels of government too?

Carbon Markets

Once all the efforts to reduce the output of greenhouse gas emissions in business operations have been exhausted, the voluntary (unregulated) and compliance (regulated) carbon markets are there to enable emitters to buy permits to do so (emission trading systems, or 'allowances'), or to reduce or remove emissions (baseline-and-credit systems, or 'offsets') with exposure to projects that claim to do so.

The European Emissions Trading System (ETS) is amongst the most advanced and now the carbon border adjustment mechanism (CBAM), the first carbon border tax of its kind in the world, requires firms in Europe to pay tariffs on some carbon-intensive imports linked to the domestic carbon price under the ETS. It aims to level the playing field for European industries that already pay for their emissions via the ETS and the export of carbon emissions from production to places where it is less heavily taxed, so to speak 'carbon leakage'. The hope is that introducing this tax as a cost of accessing lucrative European markets encourages wider adoption of carbon pricing. Time will tell and there will no doubt be differences of opinion.

However, in the absence of a global carbon tax, which most economists would argue would be, by a distance, the most efficient and effective solution to charge and pay for the damage caused by carbon emissions, carbon markets step in. They turn emission reductions and removals into tradeable assets. But there again, there is room for confusion. Both allowances and offsets are traded on the basis of a tonne of CO_2. However, companies are trading allowances and permits to pollute in the future that must eventually be surrendered to the regulator when a tonne of CO_2 is released. In an offsetting mechanism, the traded emission reductions have already happened. This is why offsets are widely viewed as a zero-sum game, because 1 tonne of CO_2 is emitted in one place and another tonne is reduced somewhere else. If this is the case, they can't be used to reduce emissions in the long term and are incompatible with a transition to net zero emissions. The cost and quality of allowances and offsets varies dramatically globally, from approximately US$10 a tonne to US$100 a tonne. The voluntary carbon market for offsets has been a particular target for sceptics based on the perceived quality of some carbon offsetting projects. So, yes, enter standards. First the Taskforce on Scaling Voluntary Carbon Markets and then, superseding it, the Integrity Council for the Voluntary Carbon Market. Despite its nascence and some controversy, the voluntary carbon market is tipped to grow from US$300 million to US$50 billion in the coming years. In the meantime, many organizations are going back to natural capital, to plant trees and mangrove swamps, to conserve and restore peatlands, to cultivate kelp and to support re-wilding projects to remove and sequester carbon.

The Herd

For better or worse, markets may be relatively efficient in the long term but are susceptible to bouts of fear and greed in the short term. They can think like an organism, often as a herd, but they can also face intransigence and institutional inertia. They like asset classes and categories, and students of markets like taxonomies. Values in markets are often driven by weight of capital as much as intrinsic worth. Markets fundamentally involve competition and are fuelled by media and culture. These are all features of the rise of the ESG and sustainable finance markets that I have witnessed and been part of for the last 15 years.

And like other aspects of society, politics and the economy, markets move in cycles and can appear to repeat history. As an example, they also repeatedly get very excited, and desperately disappointed, about innovation and technology.

- As the *Harvard Business Review* and Bloomberg recently put it, ESG ratings 'don't measure a company's impact on the Earth and Society. In fact, they gauge the opposite: the potential impact of the world on the company and its shareholders'.
- While well-meaning, some ESG standard setting runs the risk of promotion of 'paper decarbonization', whereby investors avoid high emitters to get a green label rather than engage in real-world efforts to reduce carbon.
- Investors should be demanding to know how much waste is being generated, at what cost, and what is being done about it.

CHAPTER 10

Shaping the Future Through Innovation and R&D

- While we have many of the solutions to today's environmental problems, innovation can drive transformational change.
- Research and development have been lagging in the West and accelerating in the East.
- New frontiers are being created by computing power and in space.

What do we know now that we didn't know 15 years ago? I have a fascinating benchmark. High up on my bookshelf at home I have a set of cards from 2007 called 'Drivers of Change', produced by a consulting engineering firm. The set, reproduced to this day, covers a range of themes, such as energy, water, waste, climate change and urbanization.[1] Opening the energy section, you may expect to be looking into the ancient history of clean energy technology. But no. First theme up, the 'hydrogen economy', with a picture of a zero-emission bus that I could have sworn I saw in a 'future of transport' summit last week. Next up, demand management, carbon sequestration, CO_2 storage, nuclear waste, biofuel, stranded assets, micro-generation, photovoltaics, and wind farms. The pack includes issues, some of the burning platforms of the day, such as inflation, gas imports, emissions, fuel poverty and energy access. So far, this pack could have been produced today. Only the topics of 'peak oil', 'stranded assets' and 'austerity' feel a little like yesterday's arguments, although you do get the haunting feeling that we will see them again tomorrow. Ok, some things have changed 15 years later. In the updated set, there are new sections on 'grid parity renewables', but not new renewables, and a companion section on 'electricity storage' was new. There is a new section on air quality. 'Peak oil' has changed to 'unconventional hydrocarbons' and 'extreme production'. The rest is substantially the same, save for labels. 'Demand management' has changed to 'dynamic demand'. 'Micro-generation' has become 'distributed generation'. 'Gas imports' has changed, slightly ominously, to 'gas security'.

What do we conclude? Costs have reduced for renewables and enabling solutions have since emerged to help integrate them. Urban air pollution has got worse. And as for gas? Well, that is probably the most astonishing part, and an astonishing illustration of how much, or how little, has changed. This is what the section on gas imports said in the 2007 set: 'Most Central and Eastern European countries are heavily dependent on Russian gas. A non-temporary shortage could throw many of these countries into considerable difficulty both in respect of industrial production and in terms of the comfort and safety of their peoples. The dispute between Russia and the Ukraine in January 2006 over gas prices, and the subsequent temporary cut off in supplies, caused political shockwaves across Europe. It raised the question of whether Russia may cut off gas supplies in the future for political reasons.'[2] In so many ways, we have known the problems and solutions for a long time. The biggest innovation would be doing something about it.

How

We have most of the technology today to fix most of the problems. Being resource efficient, achieving sustainable development, does not rely on technologies yet to be invented. It even costs less to use less, and is more productive and therefore more profitable if we do more with the same. The question is 'how'. So, we have two challenges with innovation. First and in the short term, it is clear that what we are doing and how we are doing it is not working, so we need to change, which will require rapid innovation in application and business models. Second and in the longer term, because we now have a much better idea about what we don't know, we know what kind of new tools and technology we need to develop and, intriguingly, the technology itself will probably help us innovate faster.

Today's technology can already make step changes in resource use. This will be achieved not only through centralized, utility-scale clean energy supply that adds energy to the system, but also by solutions that reduce demand, primarily by improving productivity. Decentralization and efficient use of established technologies such as cogeneration, on-site solar and geothermal, district energy and heat pumps, electric and thermal storage, LED, replacement of F-gases in refrigerants, smart controls, and motors. The problem is not the solutions, we know most of them, but improving and applying them. This requires effort and innovation by doing things differently. Different decisions need to be made at every step of the supply chain. Here, demand signals are the most powerful driver of change. And we need to research and develop the most efficient and productive technologies and solutions to meet these needs.

An intriguing conundrum has been that of nuclear. In principle, nuclear generators can be built at large scale with existing technology. They can produce energy with low levels of CO_2, competitive with renewables. They can do a different job to renewables, given that they can operate at efficiency levels of over 90 per cent, be available, to all intents and purposes, all the time and can produce heat as well as electricity, as well as be used to generate hydrogen. Pressurized water reactor (PWR) technology constitutes most the world's nuclear power plants. Waste management and decommissioning are planned and priced in from the outset. Nonetheless, achieving either scale or replication has proven frustratingly elusive. Cost overruns of five times and delays of a decade at Flamanville in France are mirrored in the UK, Finland and even China. It takes at least a decade to build a nuclear power plant and the costs appear substantial compared to other forms of clean generation. Why? Projects are often delivered as a one-off, constraining the development of a supply chain or the enablement of learnings to bring down the time and cost of future construction. Doing matters. If newer techniques such as small modular reactors, some of which involve recycling of waste, are to succeed, then they will require investment of a combination of learning and time. The Lawrence Livermore National Laboratory announced the milestone achievement of fusion ignition in December 2022. Albeit at very small scale, for the first time it demonstrated the potential to recreate the conditions in the centre of the sun to generate abundant clean power. It will take decades to achieve efficiencies and to apply it, but it demonstrates the power of research and development.

Research and Development

In 2021, the International Energy Agency released its 'Net zero by 2050' report.[3] While setting out a roadmap, it pointed to a gap in innovation and research and development. It said that technologies in the market today can provide nearly all the emissions reductions needed for 2030, but only 50 per cent of what's needed for net zero by 2050. The other 50 per cent of the technologies were at piloting phase. This was particularly the case for solutions for heavy industry and long-distance transport. Examples of technologies highlighted included carbon capture utilization and storage in cement, ammonia-fuelled ships, hydrogen-based steel production, direct air capture, refrigerant-free cooling and solid-state batteries. Innovation cycles need to be at least 20 per cent faster than the fastest energy technology developments in the past, and 40 per cent faster than solar so far.

The pace of innovation depends, to some extent, on picking which tools we are going to use. Reducing production costs and the development of dedicated enabling infrastructure and supply chain will be critical to scaling up, as it has been for wind and solar, and by contrast as it has not been for nuclear. Understanding resource scarcities in the supply chain will also be critical, not just to enable deployment but to find solutions to the scarcity itself. In the world of motors, through which half of the world's electricity passes, some of the breakthrough innovation has come not from making a better (i.e. new) 'mousetrap' but from making the existing mousetrap better.

Switch-reluctance technology (which I particularly like because it works based on electromagnetic 'Maxwell' force) allows manufacture without dependence on hard-to-source and environmentally damaging rare earth supply chains. In the world of hydrogen, some developing electrolysis technologies, such as solid oxide cells and anion exchange membranes, also rely less on critical minerals. Battery and solar PV may be more developed, but emerging technologies could transform their ability to scale up based on improved cost and efficiency, for instance through recycling and reduced consumption of raw materials. For example, emerging sodium-ion battery technologies benefit from abundant and cheap minerals. Solid-state batteries offer improved performance. Dry-film battery production coats the electrodes of energy storage cells instead of using liquid chemicals, which saves energy and eliminates toxic solvents. Room temperature superconductors might revolutionize electricity transmission and storage. Commercial-scale heat storage, carbon capture and utilization, industrial-use hydrogen and green ammonia production (for fertilizer) might have profound effects on the productivity of both conventional and renewable energy systems. Organic, and non-silicon thin-film, solar PV technologies might improve efficiencies and reduce manufacturing costs.

Meanwhile, government spending on R&D had fallen as a share of GDP from a peak of around 0.1 per cent in 1980 to around 0.04 per cent in 2021. Global public R&D spending on all types of energy is estimated to have been around US$38 billion, the overwhelming majority of which went to clean energy technology, half of which went to energy efficiency and nuclear. The IEA estimates that US$90 billion a year will be needed by 2026 to bring demonstration projects through in time to meet a net zero by 2050 target. The USA appears to be stepping up. The 2021 Bipartisan Infrastructure Law and the 2020 Energy Act provide US$62 billion of funding for major new clean energy demonstration and deployment programmes, tripling the previous R&D budget. Europe has also stepped up with €5.4 billion to fund a hydrogen project of 'common European interest' in the face of the energy security crisis brought on since Russia's invasion of

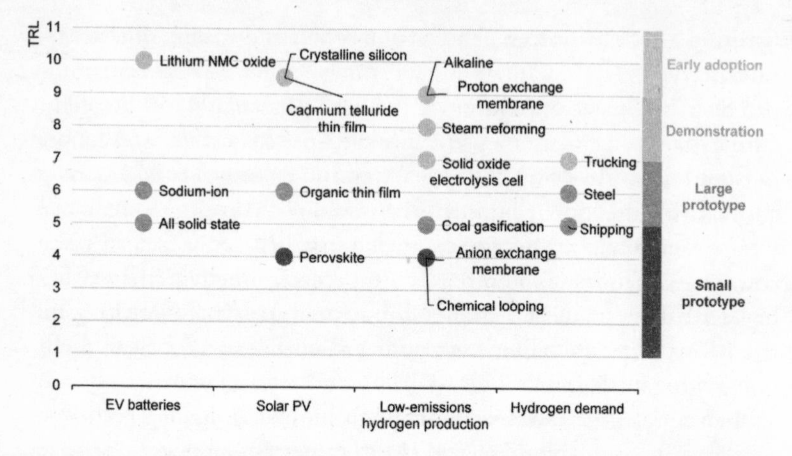

Figure 10.1 Technology readiness of solar PV, EV batteries and hydrogen technologies

Source: IEA (2022) 'Securing clean energy technology supply chains', https://iea.blob.core.windows.net/assets/0fe16228-521a-43d9-8da6-bbf08cc9f2b4/SecuringCleanEnergyTechnologySupplyChains.pdf

Ukraine. The 'IPCEI Hy2Tech' project is hoping to attract another €8.8 billion in private investment. It will involve 41 projects focusing on generation, fuel cells, storage, transportation, distribution and end-use, particularly in transport.[4]

China's Leapfrog

Leadership in clean technology development and investment was, in fact, taken by China a decade ago. Just as it took dominance over the solar industry through panel production, securing an 80 per cent market share by 2019, so too it has anticipated and secured its dominance in new markets. Factories in China now account for nearly three-quarters of global EV battery production. China has secured a 90 per cent market share for processing rare earth elements – the oxides, metals and magnets used in batteries. More than 90 per cent of the world's battery-grade lithium is produced from refineries in

China, which also process most of the cobalt and nickel, other critical battery materials. China's cost advantage is up to 40 per cent compared to Japanese competitors, and is set to increase to 50 per cent. China's capital markets are providing the financial capital and eclipsing the rest of the world. The USA and partners such as Australia and South Korea spent hundreds of millions of dollars to challenge Chinese supremacy. Meanwhile, in the first half of 2022, just three companies, China's Contemporary Amperex Technology (the world's biggest battery maker) Tianqi Lithium and Huayou Cobalt, were raising over US$10 billion to secure a further lead over clean technology supply chains.

In China, innovation is mandated from the top down and a large part is clear for the world to see. In 2006, the Chinese government's 'Medium-to-Long-Term Plan for the Development of Science and Technology' (MLP) laid out its objective to transform China into 'an innovative society' by 2020 and a world leader in science and technology by 2050. It goes back further to the 1980s and 1990s when China created the National Natural Science Foundation and the State Key Laboratory programme, and then revamped what had been a Soviet-style Chinese Academy of Sciences so that it could fund peer-reviewed pre-commercial university research in a similar way to the National Science Foundation in the USA. It then partnered with regional governments to finance high-tech zones to commercialize innovation. Shenzhen, now a city of over 17 million people (much more than London or New York, and China's fourth most populous city) with a GDP of nearly US$500 billion (more than Ireland) and a special economic zone, was the first such high-tech zone to be developed in 1985. (My firm, Sustainable Development Capital, has maintained an office and a registered energy services company in Shenzhen for the last decade, looking forward to the opportunity to deploy it at scale.)

The Chinese government's ability to capture and leverage innovation is manifest in its success in the wind turbine industry. The government launched an open bidding process for wind farm projects in 2002 to encourage competition amongst turbine makers. This opened the floodgates to foreign imports. The government then required state-owned enterprises to source 70 per cent of their components from domestic firms. Foreign firms continued to invest directly in China, but by 2009 six of the top 10 wind turbine firms were Chinese, a position that China has held on to ever since.[5] The 2006 MLP aimed to reduce China's reliance on imported technology to 30 per cent, to increase domestic R&D funding and to leapfrog international competitors in 'strategic emerging sectors' such as energy efficient technologies, equipment manufacturing, information technology and advanced materials. So, the government used export subsidies to improve the competitiveness of Chinese firms and introduced policies to require government ministries and state-owned businesses to procure goods from Chinese companies. Complaints that these measures breached World Trade Organization terms did little at first to deter foreign R&D centres being established in China, or trade. But after the US solar PV manufacturing industry was destroyed by lower-cost imports from China, the USA imposed tariffs and restrictions on Chinese-sourced equipment. In June 2022, the Biden–Harris administration invoked the Defense Production Act to ramp up 'Made-in-America' solar PV deployment in the context of the energy security crisis following Russia's invasion of Ukraine. To deal with the time that it would take for domestic production to scale up, it introduced a 24-month suspension of import duties from Cambodia, Malaysia, Thailand and Vietnam, but not from China.

Tensions with the West have come through other expressions of China's appetite to progress and innovate, which has taken many forms, from securing resources, to renting foreign talent, to acquiring technology, to being accused of stealing it. Intellectual property rights have long been high on the diplomatic agenda, alongside human rights and a trade war turned 'cold war'. The Chinese 'Belt and Road Initiative', a history-invoking centrepiece of Chinese foreign policy, is a global infrastructure development strategy launched in 2013 that was incorporated into the Constitution of China in 2017 to invest in nearly 70 countries and international organizations. (President Xi Jinping originally announced the strategy as the 'Silk Road Economic Belt' in a visit to Kazakhstan in September 2013, referring to historical land and maritime trade routes.) By March 2022, 146 countries had signed up. The initiative covers an estimated 60 per cent of the world's population, some 35 per cent of today's global economy and what is projected soon to be 40 per cent of total world trade.. One key objective is to integrate commodities-rich countries more closely with the Chinese economy. Another objective is to establish a network of worldwide research activities.[6]

Back in the UK, universities such as Oxford and Cambridge depend on Chinese student income. China has invested £134 billion in over 200 UK companies and acquired stakes in nuclear power plants, battery makers, the distribution arm of the National Grid, Thames Water, Heathrow Airport, pizza restaurants, travel agents, football clubs and even the company that builds London's iconic black cabs. Concerns around intellectual property and security abound. The Chinese Communist Party requires a representative to be present in every domestic company with more than 50 employees, while every company with more than 100 employees must have a party cell with a leader who reports directly to the party in the municipality or province. The influence of the state is felt overseas. The USA's Office of the National Counterintelligence Executive has described the Chinese as 'the wor's';s most active and persistent perpetrators of economic espionage', while the FBI has claimed that China has around 30,000 military 'cyberspies' and 150,000 hackers to target technologies in which China wishes to lead the world.

When I was in China in 2006 at the time that the MLP was launched, the frustration was described like this: 'things are made in China, not by China.' China wants to change this. (On one of my first visits to the Chinese mainland, I checked into my hotel and turned on the TV while I was unpacking. I almost always go straight to CNN. A segment covering China's struggle to innovate started up. A minute or so into the segment, the screen went blank. By the time it came back on, I had sadly missed it.)

Eco-Cities

My own journey in sustainable development financing and innovation started in China in 2006. It was five years after China joined the World Trade Organization. GDP growth was averaging 8 per cent per annum and hundreds of millions of people were being brought out of poverty, in part thanks to a massive urbanization programme. Urbanization caused its own problems, particularly in terms of resource consumption and the Ministry of Housing and Urban Rural Development wished to do something about it. The concept of the 'eco-city' was born, and one of the first sites was going to be Dongtan Eco-City on the sandy eastern sea-banks of the Jiangsu estuary, on the coast of Shanghai. It was going to be the first to market as a truly sustainable city, powered and fuelled by renewable energy and, it was hoped, a forerunner of Abu Dhabi's planned experiment in the desert, Masdars and the Saudi Arabian 'The Line' project which broke ground in October 2022. I was at HSBC and in the process of building a real estate fund for China. We were asked to invest.

At the time, the phenomenon of ghost town developments and infrastructure assets was becoming notorious in China and was cause for caution. China had built the fastest train in the world, the Maglev, to nowhere. When I took it, the thrill of travelling at 400 kmph was palpable but it was, quite literally, going nowhere fast. You got off in a random suburb of Shanghai, having come from the airport, and then got a taxi the rest of the way, wherever you were going. Ghost towns (my name for them) had been built speculatively in Shanghai outskirts as satellite towns between 2001 and 2006, as part of its 10th Five-Year Plan (2001–2005), the principal slogan of which was 'urbanizing the suburbs', and the 'One City, Nine Towns' pilot projects. They often had themes. 'Thames Town', for example, was completed in 2006.[7] It was meant to house up to 10,000 residents, but remained mostly empty, with most properties acquired as second homes rather than reducing overcrowding and congestion in favour of garden cities, as the government originally intended. Eight other towns were built around European and American building styles.

My proposal for Dongtan Eco-City was to do things differently. To start with, the idea was to create something for people to do when they got there, or even better, a reason that they would be excited to go and stay there. The idea, which aligned with Chinese government ambitions, was to create, so to speak, the Harvard and MIT of environmental economics and technology. Modelled on the success stories of the Silicon Fen in Cambridge, UK, the cluster of technology and financial services businesses in Cambridge, Massachusetts, and underpinned by the multi-disciplinary basic research model so effectively and profitably delivered by the Weizmann Institute of Science in Rehovot in Israel, Dongtan would attract an industrial cluster around the economic draw of a major educational institution of global relevance with innovation at its core.

The city would be designed with state-of-the-art energy efficient methods, the most integrated and resource efficient urban masterplan ever created, and be energized by clean and renewable technology, a showcase for the modern era of sustainable development. It would be delivered and financed through an international public – private partnership. Both Cambridge University's Programme for Sustainability Leadership[8] and the Weizmann Institute agreed to participate, modelling the curriculum on their unique insights and experiences. The project, the 'Institute for Sustainability', became a cornerstone of the British government's UK – China Sustainable Development Dialogue, coordinated by Number 10 and the Foreign Office. It was to be built in China and replicated in the UK.

In January 2008, three months after launching Sustainable Development Capital, I co-signed a memorandum of understanding, witnessed by the UK Prime Minister, Gordon Brown, in Shanghai, to deliver the project. The Shanghai and Beijing governments both got behind the plan. I remember feeling like the world was turning on its axis, that we were going to make a real difference. In the summer that year, the Shanghai party secretary who had supported the project was jailed. The lights went out on the project as the doors closed behind him.

Ten years later, at the Goldman Sachs Environmental Innovation Conference in New York, I met an executive from Sidewalk Labs, the urban innovation arm of Alphabet (then Google). The concept of the eco-city had arrived in North America, US style. The language had changed, and it was based as much on software as hardware. The new language was a 'smart city' and the exceedingly cool Sidewalk Labs had won a bid, in October 2017, to redevelop a 12-acre site, Quayside, in Toronto. The development was going to be a showcase for the optimized urban experience. Autonomous taxis, heated pavements and automated refuse collection would be complemented with digital technology so that everything could be monitored with sensors, whether you were crossing the street or relaxing on a park bench. The development was meant to be built 'from the internet up'. (I often find myself thinking of two Hanna-Barbera productions that I watched as a child. The Flintstones lived in the Stone Age. The Jetsons lived in a city of the future, getting to work in flying cars and taxis, and every whim was served by robots of one description or another. I was captivated. I have never decided which version of the world I like best, and I am delighted that in that respect, I never have to choose.)

But in May 2020, Sidewalk Labs aborted the project two weeks before the development agency, Waterfront Toronto, was scheduled to vote as to whether to shut it down. They blamed COVID-19, but truth was, the people of Toronto didn't want the routine activities of their lives to be monitored, let alone by the private sector accused of 'hubris'. The idea simply didn't chime with civic culture. In 2022, the Waterfront Toronto project has been re-drawn with nature at its heart, citing the 'importance of human life, plant life and the natural world', including a 2-acre forest and a rooftop farm.

Compute

Can new technology help us innovate faster, an auto-catalytic virtuous circle of improvement, invention and reinvention? Beyond the ability of human intelligence to innovate, we are working on developing artificial intelligence or 'AI', machine learning. AI could help strengthen climate predictions, enable smarter decision-making for decarbonizing industries from building to transport, and work out how to allocate renewable energy. Google's research lab, DeepMind, allowed its AI system to control temperature inside its data centres, helping to cut energy bills. Similar technologies promise to help optimize smart energy management control energy demand in buildings and industry.

Digitization will make a substantial contribution to measurement and energy management and, therefore, the ability to control outcomes in buildings and industry by avoiding needless waste of energy. It is a welcome companion to physical equipment, configuration and process changes that can make a substantial difference. Pair digital controls with interventions that can affect step changes in consumption, like liquid cooling solutions for datacentres that can cool data racks rather than the room, slashing energy demand, or new building construction methods and materials that can produce low-cost zero carbon buildings, and real change happens.

Beyond building automation and controls, there are, however, constraints and limits. Some are ethical. Machine learning applications have raised concerns about creeping public surveillance, intentional misuse, privacy, transparency and data bias that can lead to discrimination and inequality. The answers delivered by machine learning also depend on the questions asked. But with what consequences? We face these conundrums every day. Solve climate change but with what impact on the environment, natural resources, and society? Other constraints lead us back to energy. A University of Massachusetts study estimated that training a large AI model to handle human language can lead to emissions of nearly 300,000 kilograms of carbon dioxide equivalent, about five times the emissions of the average car in the US, including its manufacture.[9] Datacentres have been estimated to use some 2 per cent of the electricity in the United States,[10] but some research has proposed that the energy needed to process AI computations could increase this share to 10% of global electricity within the 2020s.[11] With the advent of AI systems such as ChatGPT, this may, in due course, turn out to be an underestimate.

Space

Thanks to innovation, technology and R&D, we can derive information from mechanical, electrical and digital infrastructure and even conduct preventative maintenance, to all intents and purposes anywhere and anytime. This is thanks to a massive physical, global, terrestrial web of fibreoptic and datacentre infrastructure empowered by increasing amounts of energy, wherever and for as long as, and at whatever price financially and environmentally, it is available. It is also thanks to the innovation that opened space as a new frontier for communication, but also potentially for conflict.

Those of us who grew up in the post-war twentieth century were born into a space race between the USA and the Soviet Union. The technology behind it originated primarily from military and national security objectives. The ensuing competition gave us the satellite communications technologies that all civilians rely on today for instantaneous voice and video communication, navigation and positioning, financial transactions and, regarding climate and the environment, monitoring, measurement and prediction of the weather (which supports agriculture, farming and fisheries), fires and greenhouse gas emissions.

Now space has new competitors. Enter China and India and the role of the private sector. Increasing competition for finite resources, even in space, such as low earth orbits and access to spectrum, will create tensions between nation states and other competitors. The space environment is also experiencing its own environmental limits when it comes to waste. Low earth orbit (LEO) is anywhere from 100 kilometres above the Earth to around 36,000 kilometres above the Earth. Beyond that point is cislunar space, where geostationary orbit accommodates national security and some commercial satellites. This is regulated by the International Telecommunication Union. LEO, though, is where the satellites are for weather, navigation, and internet. It is not regulated, and spectrum or orbital slots only require a licence from the national space agency, for instance from the Federal Communications Commission in the USA. As such, orbital slots are awarded on a first come, first served basis.

Today's 4,550 satellites (more than half of which are inactive or dead) in LEO are due to be joined by around 13,000 registered by China alone by 2030. India wants to launch thousands more. Private companies such as SpaceX and Amazon are planning to launch thousands of satellites to achieve global satellite internet coverage. All in all, this could deliver another 50,000 satellites into orbit. Limited prime sites in LEO provide the opportunity for conflict, and accidents. Environmental issues are also ramping up. Trackable orbital debris – or space junk – has increased by 80 per cent in the last two decades. It is estimated that there are now over 34,000 pieces of space junk bigger than 10 centimetres and millions of smaller pieces. Eventually they will come down to Earth, but even in the meantime, if they hit something, it could be catastrophic. Satellites and the International Space Station all perform collision avoidance manoeuvres. Collisions are rare. A Chinese satellite broke up in March 2021 after a collision with a Russian rocket, and the last satellite to be destroyed by space junk before that was in 2009. But the risk is on the rise.

While most space debris lands in the oceans, some does find its way onto land. In July 2022, an Australian farmer discovered a piece of a SpaceX capsule sticking out of the ground on his land. The only recorded case of someone being hit was in Oklahoma, in the USA, when a piece of metal knocked the shoulder of Lottie Williams as she walked in the park in Tulsa. Other incidents included damage to buildings in the Ivory Coast in 2020 after they were hit by pieces of a Chinese rocket. A study by the University of British Columbia rates the chances of someone on Earth being killed by space debris in the next decade as 10 per cent.[12]

Steps have been taken to improve resource efficiency in space. SpaceX, for example, has designed its rockets to be 'fully re-useable', whereas rockets are traditionally discarded after launch.[13]

The next frontier, though, is new resources. China entered the space race and is only the third country in history to have both put astronauts into space and built a space station. NASA, the European Space Agency, the Japan Aerospace Exploration Agency, and the Canadian Space Agency are collaborating to place a space station in lunar orbit, the 'Lunar Orbital Platform-Gateway'. It is planned to serve as the staging point for robotic and crewed exploration of the lunar south pole and is the proposed staging point for NASA's 'Deep Space Transport' plan for transit to Mars. Launch is planned for November 2024.

In 2021, China put the first module of its Tiangong space station into orbit. Chinese astronauts are excluded from the International Space Station because US law bans its space agency, NASA, from sharing data with China. In July 2022, Russia announced that it would withdraw from the International Space Station, which has been in orbit since 1998, after 2024 and build its own space station instead. Relations have soured since Russia's invasion of Ukraine. The European Space Agency (ESA) had, by then, already pulled out of its deal with Russia's Roscosmos agency to launch a rover to Mars, and Russia had stopped sending its Soyuz spacecraft from an ESA launch site in French Guiana.

By 2030 China wants to have landed its first astronauts on the Moon. Why? NASA reports detection of basalt, iron, quartz and silicon, common resources that could be turned into buildings, windows, stoneware and solar panels. But there are more scarce materials and precious metals too, such as platinum, palladium and rhodium, used in electronics, and titanium ore, potentially ten times richer than that found on Earth, which mixes with aluminium or iron to make an alloy that is lightweight, corrosion-resistant, strong and resistant to extreme temperatures, that could be used to build engines, medical implants and structural frames. There are 17 rare metals that are scarce on Earth, including scandium and yttrium, which could be used in vehicle engines, to make glass or ceramics, electronic devices, radar systems and superconductors and more. Rarer still is helium-3, a gas that could be used as a clean and powerful fuel for nuclear fusion reactors.[14]

So what? The UN Outer Space Treaty of 1967 states that nowhere in space can be claimed by any one nation. The UN Moon Agreement of 1979 states that space should not be commercially exploited. The USA, China and Russia have refused to sign. The USA is now promoting its Artemis Accords, spelling out how nations can exploit the Moon's minerals in a 'co-operative way'. However, Russia and China refuse to sign the Accords, insisting that the USA has no right to make the rules for space.[15] Resource wars could become star wars.

Science and Policy

Back on Earth, a crucial role for R&D is the interaction of science and policy. During the COVID-19 pandemic, we all got used to a different lexicon. For a while I kept a list of phrases that entered the popular dialect, some of which we may never use again (we hope), such as 'flatten the curve', 'social distancing', 'wash your hands', 'lockdown' and 'PPE'. However, one of the key fundamental tenets was this: 'follow the science'. The UK government, infamous for having said 'the people of this country have had enough of experts' when debating the country's future in or out of Europe in 2016, laid its policies and the operations of all business and society in the hands of scientists, under the Chief Medical Officer and the Chief Scientific Officer. As Corinne Le Quéré, former lead author for the UN IPCC and current member of the UK's Climate Change Committee and Chair of France's High Council on Climate, put it at the time: 'Look at the size of the actions we're taking [against COVID-19], you need to respond to climate change at that size.'[16]

In the crises of the climate and environment, we have a pillar scientific body, the UN IPCC. It published the first part of its Sixth Assessment Report, 'The physical science basis', in August 2021. It said: 'It is unequivocal that human influence has warmed the atmosphere, ocean and land.'[17] It published the second part, 'Impacts, adaptation and vulnerability', in February 2022. It said: 'The rise in weather and climate extremes has led to some irreversible impacts as natural and human systems are pushed beyond their ability to adapt.'[18] It published the third part on 'Mitigation of climate change' in April 2022. It said that the current Nationally Determined Contributions (NDCs) for greenhouse gas emissions would likely result in warming exceeding 1.5°C by 2030. To limit warming below 2°C, significant and rapid mitigation efforts must be accelerated after 2030.[19]

While there is ample scope for debate as to the range of plausible scenarios, degrees of warming and quite how much damage will be done to humans and nature, the science and the overall messages it offers appear clear and in plain sight. However, science collides with policy, politics and business on the one hand, and media and communications on the other.

- Technologies in the market today can provide nearly all the emissions reductions needed for 2030, but only 50 per cent of what's needed for net zero by 2050.
- China is leading the charge.
- Competition for resources is finding new limits.

CHAPTER 11

Talking to Ourselves

MEDIA AND COMMUNICATIONS

• • •

- Climate change is a defining issue of our time but divides opinion.
- Polarization is manifest along political lines and in the messages from media.

The best ideas and technologies don't always win. More than 4,000 mousetrap patents exist, yet only around 20 ever became profitable products.[1] Enthusiasm for and investment in clean technology and solutions, and even more fundamentally, dealing with climate change and the environment, will go through waves, with highs and lows. The tides will be affected by politics, capital markets, commentators and educators. The solutions that succeed will do so based on productivity, profitability, persuasion and policy. This is when communication and media take centre stage. And they will go through waves too, as they are today.

Polarization

The United Nations has described climate change as 'the defining issue of our time'.[2] Yet, or more likely as a result, it is highly politicized, commands widespread media coverage, charges debate and provokes extreme positions to be taken, from atavistic denial on the one hand to alarmist hysteria on the other. Where are the barricades and why are they there?

It is easy to depict and tempting to accept a fundamental divide down the middle of the political spectrum, where the left 'believes' in climate change or thinks that it is serious and requires urgent action to deal with it, whereas the right doesn't, or doesn't as much. The evidence appears to support this. A recent survey by Reuters and Oxford University[3] suggests that climate scepticism is highest on the right of US politics, where only 18 per cent of those that place themselves on the right consider that climate change is very or extremely serious, compared to 89 per cent of those on the left.[4] A 2020 survey by Pew Research Center showed that 72 per cent of Democrats thought that humans cause climate change, compared to only 22 per cent of Republicans.[5] However, taking a more global perspective, the numbers even out a bit more, with 58 per cent on the right and 81 per cent on the left. The number in the centre is 71 per cent. Nearly two-thirds (63 per cent) of the 66,445 global respondents placed themselves in the centre, as did nearly half of the respondents from the USA. While the overall message is that most people think that climate change is 'very or extremely' serious, nonetheless in countries such as the USA, where considerably more respondents identify with the right or the left than the global average, the issue of climate change seems highly politicized. This is reflected in media and impacts policy.

Tone

However, it also turns out that the tone matters. Other polling results, for example, showed that US Republicans responded better to messages of opportunity rather than fear, particularly starting with benefits to the individual, family, neighbourhood, community and then country, in that order.[6] Themes that resonated particularly well with Republicans included creating a 'cleaner, safer and healthier' environment, which elicited double the level of positive voter response to the phrase 'sustainability'. Indeed, this theme appeared to cut across the political spectrum. When asked, 75 per cent of US voters considered that it was more important to protect 'the economy' than the 'environment'. However, when the question was rephrased to offer the choice between 'the economy' and 'a healthy, safe, clean environment', 55 per cent chose the latter. By contrast, the polling suggested that vote-losing phrases in the USA included 'net zero', 'carbon emissions', 'greenhouse gases' and 'global problem'. Supporting this argument, other polls found that Democrats and Republicans were almost completely aligned on ideas such as 'planting about a trillion trees to absorb carbon emissions', but far apart on ideas such as 'tougher' restrictions on power plant carbon emissions and fuel efficiency standards for cars, and 'taxing' emissions.[7] (In my own experience, messages such as 'cheaper, cleaner and more reliable energy', improving 'resilience' and energy 'security', 'cutting costs' and 'improving productivity', 'profitability' and 'growth' through energy efficiency, 'doing more with less' seem to cut through in our capital raising numbers, as well as, crucially, and a sine qua non, 'investment performance'.)

Opinion Forming

We need to look at how attitudes are formed and the media that shape them. We can start with where news and information come from. Overall, people pay more attention to the television than any other media, by far – some 35 per cent compared to the next biggest media, online news sites from major news organizations at 15 per cent. Printed newspapers and radio (5 per cent) are each less influential than social media (9 per cent) or even conversations with friends and family (6 per cent). The influence of the internet and social media is higher among younger generations. Television remains the single largest source of news, but social media, blogs and posts, online news sites and specialized outlets covering climate issues collectively have a deeper reach. The younger generations, who watch less television, are less optimistic than the older ones. Over 60 per cent of people over 55 surveyed in the UK thought it was not too late to do something about climate change, compared to only 34 per cent of 18–34-year-olds.

Indeed, so entrenched has been the pessimism among younger generations that it has become a source of severe anxiety and a matter of mental health. 'Eco-anxiety' or 'eco-distress' was first defined by the American Psychiatric Association as a 'chronic fear of environmental doom'. A survey of 10,000 16–25-year-olds in ten countries, conducted by the University of Bath and published in 2021 showed almost half of respondents to be anxious to the point that it negatively affected their day-to-day lives, while 75 per cent felt frightened.[8]

How You Say It

Putting the pieces together, one of the pictures that starts to emerge is that news about the environment and climate change can tend to be communicated to those who want to hear it in language that they want to hear, particularly in polarized societies. In the USA, for example, around 40 per cent of people who either think that climate change is 'extremely serious' or 'not serious at all' share news about it online, in each case the strong view, compared to around 25 per cent who have less strong views. The biggest noises online are therefore made by highly vocal minorities. Media campaigns launched by the likes of the Global Climate Coalition and Citizens for a Sound Economy, backed by oil and gas companies, were highly effective because they were well funded and tapped into fears of taxes and hope that climate change was nothing to fear. Is the moral of the story that the volume needs to be turned up at just the right tone so that the largest constituency in the centre can hear and wants to listen?

An analysis of the coverage of climate change in news media globally between 2006 and 2018 (pre-pandemic), released in 2021, illustrated 'considerable, but no longer increasing attention to climate change'.[9] It will be interesting to see how coverage correlates to the extreme weather events of 2021, and the heatwaves and 'fire weather' of 2022. Coverage often focused on the societal dimension of the climate crisis, particularly in the Global South, which was more focused on impacts on humans than the climate in general, as it tends to be in the Global North. This finding aligns with a heightened concern about the serious effects of climate change in the Global South (most affected) versus the Global North identified by the Reuters and Oxford University report referenced above. People care about people most.

What You Say

Analysis of media content over longer periods dating back to the early 1990s has shown that coverage has indeed been moving from the science of climate change to its societal impacts.[10] Greenhouse gases themselves are invisible (as is wasted resource such as energy), the topic is scientifically complex and multi-disciplinary and approaches to mitigation are debated. The topic has moved from the science pages or segments to coverage of politics, the economy and society, what has been described as a 'societal turn'. The politics of efforts to reach international agreements, the ranking of climate on the issues lists at the ballot box, the capital, cost, or investment involved with the low-carbon transition, the rise of the electric vehicle entering everyday lives, or the humanitarian and security implications of resource conflicts, displacement and migration are taking centre stage because they matter to people.

Consensus

What people think matters – a lot. Attitudes towards climate change, no less than any other issue, are defined by perceived social consensus. This can be as, if not more, influential than scientific consensus, perception of which has been described as 'gateway' belief to other climate-related beliefs. A powerful example of how attitudes are formed comes from an analysis of social media,[11] from which it turns out that the social consensus largely depends, for instance, on whether blog comments endorse or reject the contents of a post. (Social media 'likes' and 'dislikes' can be relatively easily and cheaply purchased on the black market.)

When comments reject a post, perceived consensus is lower than when it's endorsed. In general, conformity of opinion tends to increase with the size of the majority. This can be rational as larger groups of people may tend to circulate more reliable information than smaller groups. When the size of the majority becomes overwhelming, then a consensus forms. ('Bots', short for 'robots', which generate online activity artificially, imitating human behaviour, can create a dangerous perception of consensus.)

The effect of the consensus can be enhanced if it involves members of a person's 'in-group', for instance family, friends, or work colleagues. Specifically in the context of attitudes to climate change, while more information has generally tended to nudge people's attitudes towards the scientific mainstream, it also turns out to be relatively fragile, perhaps because of the scale of fears and hopes involved. Research has shown that only a few dissenting views on climate change can undermine the perception of a scientific consensus even when numeric information about the extent of expert agreement is available. Indeed, a recent survey of investment managers found that they thought that only around 60 per cent of scientists believed that climate change was human-induced, some 40 per cent short of the truth.

Some powerful versions of the truth in cinema and television over the years had profound effects, such as *An Inconvenient Truth* by Al Gore or David Attenborough's documentaries. Some satire, such as *Don't Look Up*, has powerfully portrayed the propensity for denial.

In short, communicating through the media really matters. It can determine policy because it translates science into politics, drives business and establishes popular opinion. Those that want to do something about climate change, the environment and the effect on our society need to get better at it, at the same time as we make things better and do things better.

- Most people think that climate change is 'very or extremely' serious.
- Vocal minorities, particularly on the television and online, have an outsized impact.
- Protecting the interests of people and society resonates strongly and widely.
- In the USA, calls for a 'cleaner, safer and healthier' environment can unify while phrases such as 'net zero', 'carbon emissions', 'greenhouse gases' and 'global problem' can divide.

CHAPTER 12

Beyond the Edge

SUSTAINABLE GROWTH

• • •

'As the saying goes, the Stone Age did not end because we ran out of stones; we transitioned to better solutions. The same opportunity lies before us with energy efficiency and clean energy.'

Steven Chu, former USA Energy Secretary and Nobel laureate

In so many ways, we have triumphed. We have the technology to defeat viruses and extend life. We can communicate instantly in every language. We have the benefit of hindsight, with the ability to interpret our political and natural history and understand the implications of our past for the future. We can extend our reach beyond the Earth into space, analyse the gaseous planets around us and celebrate why we, amongst the planets in our solar system, have the unique conditions and resources to support intelligent life. We have even created an artificial intelligence well beyond some of our own capabilities, through machines.

Energy, industrial and technological revolutions, and progress that they have enabled, have brought profound and positive implications for quality of life and society. At the same time, we must avoid the risk of snatching defeat and disaster from the jaws of victory and triumph, as our own survival is threatened with wars fought over scarce resources. Our climate and environment have reached the brink of collapse from resource abuse and depletion, degradation, pollution and emission of greenhouse gases that could eventually make our blue Earth resemble our gaseous and lifeless planetary neighbours. Severe weather and war both generate the imperatives of energy and resource security that will force change and both conflict and cooperation.

Energy has become synonymous with climate change because it is the largest contributor to human-made greenhouse gas emissions. But it is also synonymous with security and has a defining impact on geopolitics. And it is synonymous with society, which depends on it. Because our world is being pushed to the limits over it, it is now vital and urgent that we are honest about it and do the right things about it. The transition to a lower-carbon economy is not going to be delivered by a magical transformation to wind and solar. It will not happen in time, and would, in any case, place enormous stress on the environment in terms of resource extraction and depletion. It will require the introduction of balancing technologies, some of which will be fuelled by conventional sources. The energy system is likely to get worse, in the sense of higher emissions, before it gets better. So, efficiency and productivity are key.

The journey to net zero is enormous. We are separated by a vast ocean. The pathway is unclear, and we are subject to unpredictable events on our journey, whether from the weather around us or the waves beneath us or others trying to stop us. We don't have all the solutions as we don't understand everything about our problem. The problem of our journey has been faced by many before us. Not all of them succeeded and many failed because they set off in the wrong vessel. We run all the same risks today. If we are determined to reach the other side, we can. But we need to be realistic about our journey, the vehicles we use, when we need to leave and the time that it will take. We have an incredibly long and dangerous journey ahead of us. Setting off in, or relying on, the wrong vehicles will doom it. We will need resilience and an ability to deal with uncertainty. We must change course where we need to, but not change our goal, so we don't get lost and drown on the way. For most of our fellows and passengers, our success will be measured as a pass/fail, not by how hard we tried. We must plan and choose to do nothing less than succeed.

We know a lot about what we need to do and can do it well. Doing it well will even make us more productive, healthier, and more prosperous. However, right now, there is a large gap between what we can do, based on what we know, and what we are really doing in

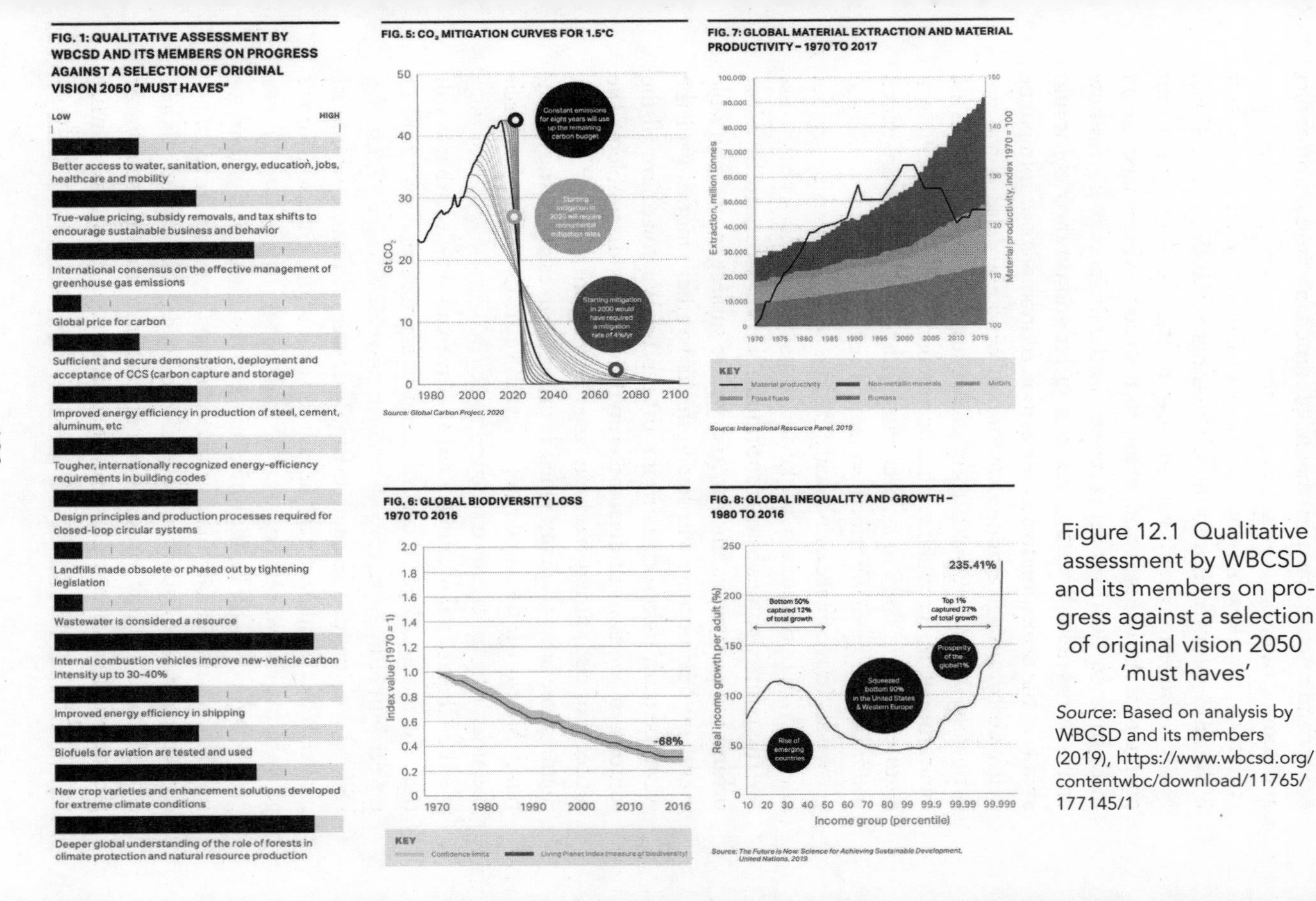

Figure 12.1 Qualitative assessment by WBCSD and its members on progress against a selection of original vision 2050 'must haves'

Source: Based on analysis by WBCSD and its members (2019), https://www.wbcsd.org/contentwbc/download/11765/177145/1

practice. We now know that net global carbon emissions from power generation, buildings, industry, and transport need to be reduced rapidly over the next 20 years, and substantially to zero over the next 30 years if global temperature rise is going to be limited to 1.5°C or 2°C.[1] We know that this is going to be hard because demand for energy generation is already set to triple by 2050. Renewable energy must triple by 2050 just to keep pace and it must displace, not just add to, fossil fuels. We also know that 'electrifying everything' would at least triple energy demand, again, so that must be built into policy and plans, so that we don't make things worse. We also know that even if we succeed in doing all of these things well, forests must act as a carbon sink. The Amazon has been absorbing 4 per cent of emissions from fossil fuels. But parts of the Brazilian Amazon are now a net carbon emitter. We have to change this.

We also have to change quickly.

The World Business Council for Sustainable Development has reviewed our progress and concluded that we are running badly behind schedule and that some things are getting worse. We have made good progress on improving the carbon efficiency of vehicles, and better understanding how important forests are, and improving farming methods to adapt for extreme climate conditions. We have made reasonable progress on agreeing that we need to manage greenhouse gases. But we are doing badly on access to essential infrastructure, efficient use of public funds, carbon pricing, managing waste and delivering on energy efficiency.

The Global Carbon Project has shown that we will have used up our remaining carbon budget in the next eight years, so we have to get on with it.

The International Resource Panel has shown that extraction of materials has skyrocketed in the last 50 years, while material productivity in the last 20 years has actually got worse.

The World Wide Fund for Nature and the Zoological Society of London has shown that biodiversity is collapsing, with nearly 70 per cent lost over the last 50 years. And the United Nations has shown that global economic growth does not translate into prosperity for all. Income inequality has got worse rather than better. Over a quarter of all the global growth of the last 40 years has been captured by the wealthiest 1 per cent of the global population. Only a little over a tenth of the growth made its way to the least wealthy 50 per cent. Zooming in on the USA and Western Europe, we find that the bottom 90 per cent of the population have actually seen their real incomes fall rather than rise.

Our current trajectory threatens a continuation of a vicious cycle of human struggle, resource scarcity, competition, displacement and loss of income and capital turning to conflict.

To turn the wheel the other way, we need to focus on productivity and efficiency, as well as real profitability. This is what is going to make growth sustainable.

Towards a Sustainable Future

Our biggest problem is that we use so much to do what we do. This is unsustainable and the planet can't cope. It's on the edge. Our biggest opportunity is to do more with less. Less energy, less natural resources, less time. This is efficiency, productivity, and is the key to sustainable growth and prosperity.

We know what we need to do and there are many millions of us who are working hard, and increasingly smart, to do it. Every day, I draw inspiration from new companies that have survived the last 15 years or are just starting today to launch products and services that are better because they are cheaper and cleaner and more reliable than the old. I am inspired by the innovators in business and finance. I am amazed by the gulf in understanding and trust between the public sector and the private sector, but see this as a huge opportunity for change, as the drive to succeed (there is no alternative for the private sector, it's sink or swim) comes together with the analytical resources and regulatory capability of the public sector.

As Hermann Hesse wrote in *The Glass Bead Game*, in the 1930s, a novel that at once explored the interrelatedness of all human knowledge and activity and, at the same time, the importance of considering what can happen next, the unknown: 'Although it is easy to fit any given segment of the past neatly and intelligibly into the patterns of world history, contemporaries are never able to see their own place

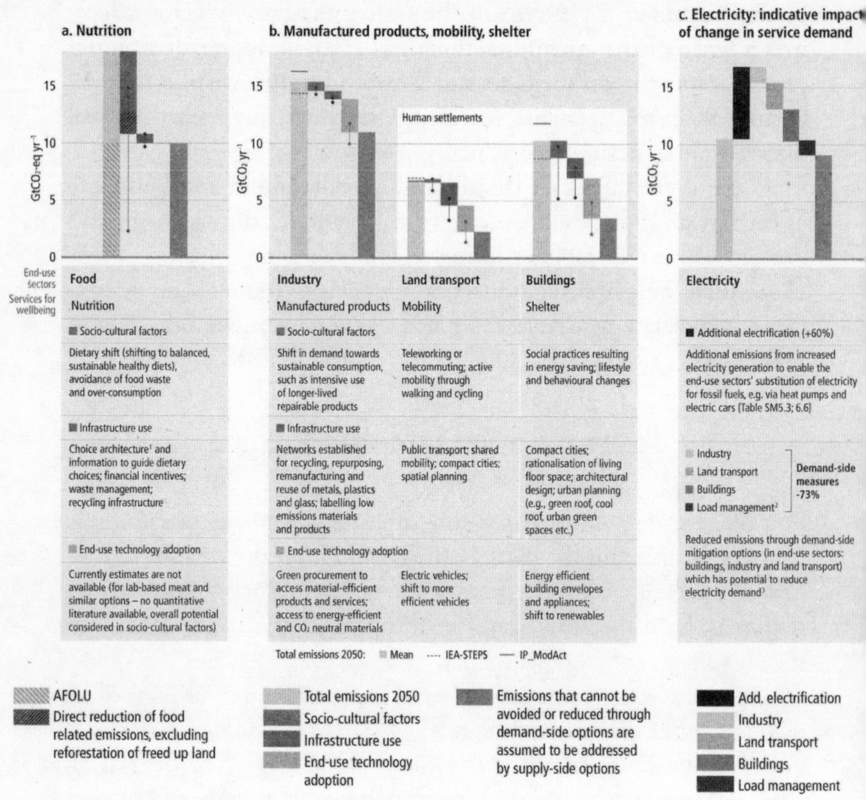

[1] The presentation of choices to consumers, and the impact of that presentation on consumer decision-making.

[2] Load management refers to demand-side flexibility that cuts across all sectors and can be achieved through incentive design like time of use pricing/monitoring by artificial intelligence, diversification of storage facilities, etc.

[3] The impact of demand-side mitigation on electricity sector emissions depends on the baseline carbon intensity of electricity supply, which is scenario dependent.

Figure 12.2 Demand-side mitigation can be achieved through changes in socio-cultural factors, infrastructure design and use, and end-use technology adoption by 2050

Source: Figure SPM.6 from IPCC, 2022: Summary for Policymakers. In: Climate Change 2022: Mitigation of Climate Change. Contribution of Working Group III to the Sixth Assessment Report of the Intergovernmental Panel on Climate Change [P.R. Shukla, J. Skea, R. Slade, A. Al Khourdajie, R. van Diemen, D. McCollum, M. Pathak, S. Some, P. Vyas, R. Fradera, M. Belkacemi, A. Hasija, G. Lisboa, S. Luz, J. Malley, (eds.)]. Cambridge University Press, Cambridge, UK and New York, NY, USA. doi: 10.1017/9781009157926.001 https://www.ipcc.ch/report/ar6/wg3/downloads/report/IPCC_AR6_WGIII_FullReport.pdf

in the patterns …' This observation applies today, where governments and societies are finding themselves caught unprepared and off guard, surprised by shocks that had roots in the recent past, as well as modern history. (Pattern recognition is the key to winning chess. There are more than twice as many chess grand-masters in Russia as there are in the USA and Ukraine, which have about twice as many as China and India.)

But we are now facing a key watershed moment in history, where choices that we make today affect the global environment and all people, not just nation states. We cannot afford to overlook our place in the patterns of geopolitics. Our choices and actions must be based on the limits of our resources and economy because they define our present and our future. There is no certainty that we will succeed in navigating the edge of environmental and social crises, conflict, or critical climate challenges, but at this stage we can still choose to do so.

Epilogue

A JOURNEY

• • •

I tried and it was hard. In 2006, while I was at HSBC, I tried to convince the bank that we should start a fund to invest in 'environmental infrastructure', including energy efficiency. The bank was a pioneer of 'social' infrastructure. The concept of investing in essential services without which society would not function, such as healthcare and education, was incredibly appealing to me and had led me to move out of a lucrative career in investment banking into investment management in 2005. HSBC had a specialist division, one of the first and only in London, which focused on public–private partnerships (PPPs) for infrastructure. It was tremendously successful and the rigour and integrity of the business in 2005 and 2006 shaped my own career and prepared me to launch SDCL, aged 33, in 2007.

My leaving served as a catalyst for change, and we were hired immediately to help design and deliver the strategy. But when it came to energy efficiency specifically, there seemed to be a scarcity of projects to invest in and the bank had other priorities than to develop them.

I started Sustainable Development Capital LLP in 2007 to create the company that I wanted to work in and to address the problems for which solutions would be most valuable and impactful. I wanted to use my place in the world, my energy, my evolving understanding of business and finance, and my persuasion, to move money into the environmental infrastructure sector. Beyond the already maturing renewable energy project sectors, I was convinced that there was an opportunity to reduce the amount of energy being used. For me, sustainable development was predicated on resource efficiency.

In 2010, we started consulting on project development. Large multinational companies, the world's largest retailers and global brands were looking to reduce carbon in their supply chains (what is now referred to as 'scope 3' emissions) because this was where some 90 per cent of the carbon impact of their products took place. We were a multi-disciplinary team, visiting and working with supply chain factories to design energy efficiency projects. Our feasibility studies were initially funded by the not-for-profit Environmental Defense Fund, which it did partly in support of US brands that were members. However, in the end, factories were reticent to share information in case their margins were squeezed. Meanwhile, brands could place objectives and targets on suppliers but could not enforce them. Why wasn't energy efficiency happening? Not because there were not savings and productivity gains to capture, but because of a lack of project development and investment. We learned a lot and our resolve that these problems could be solved strengthened.

Then, in 2011, after attending a conference in Monaco on energy efficiency, I ended up in the lift with a hotel owner. After the doors closed, I said that I could find 30 per cent energy savings in his properties. No way. He had won green tourism awards. His buildings were efficient. But he made me a bet. If I could find 30 per cent, he would do a deal with me and invest in a fund to make it happen and roll it out. I did. He did. And that's how we got started. In the basements of hotels and hospitals in London, building by building, we learned not to impose our ideas on the buildings but to allow them to communicate with us, through their engineers and through the data that was being collected every day. Although timing had been tough at the beginning, everything started to change as we discovered savings just as large as anything we had seen in China, but at least a less challenging environment in which to do business.

All this time, we had been promoting the concept of energy efficiency with the UK government. In 2010, a Conservative–Liberal coalition government had been formed, pledging to be the 'greenest government ever'. Some ideas, such as one for a 'Green Investment Bank', I had heard about when I worked on the UK China Sustainable Development Dialogue with Number 10 on the Dongtan project in China. This was now a trophy project for the new coalition government. What would it do? A core focus would be offshore wind, which made sense. The UK was home to 25 per cent of Europe's offshore wind generation resource and this offered clean energy at scale, but needed to get started. Government support could help. Waste of energy was also on the table. We were at least as keen to see energy efficiency on the table too and we wanted to do something about it.

That year, the United Nations Climate Change Conference was held in Cancún, Mexico (the Conference of the Parties, or COP16). I had been at the Copenhagen COP the year before. The political negotiation had fallen flat but I had met a lot of investors. So, a few days before, I booked the cheapest economy ticket I could find and, unaccredited, I flew to Mexico. I had a couple of days to prepare, so I used them. I organized a dinner at a decent hotel and invited a guest of honour that everyone would want to meet, the UK's new energy and climate change minister. The rest of the invitations went to investors, institutional investor groups focused on climate change and European, as well as other, international development finance institutions. We discussed the climate, and renewable energy, and better business practices and everything that everyone was talking about. But the seminal moment of truth that changed everything was a simple conversation on the balcony with the head of the European Investment Bank's climate finance division. I wanted to create the first institutional energy efficiency fund and I wanted to convince the UK government to back it. If I could, then the European Investment Bank would invest. 'We will if you will', so to speak. This was a foundation for consensus, but somebody had to move first.

There was some more distance to travel first. The Green Investment Bank had not yet been formed. In the airport waiting to fly back to London, I was approached by someone from the consultancy firm McKinsey. They were bidding for the mandate to advise the UK Treasury and department for business on the formation of the Green Investment Bank. They liked what we had been saying in Mexico and at World Economic Forum events and invited us to join a consortium with Vivid Economics (which they acquired years later) to bid. It was a great team and we won. We really did have white boards back then and we used them to allow me to outline project finance structures that we could transpose from traditional infrastructure project finance to energy efficiency project finance. We would structure project companies that would enter agreements to provide services to clients and then subcontract construction and operation to third parties that could take the risks. It was an infrastructure business model. Energy as a service.

At the same time, we were doing laps of London with private investors and small projects to prove it. It all resonated, and the government adopted it in its plans. In due course, it became clear that once the Green Investment Bank was formed, they would need managers to implement the plan. Our consortium partners gave us some valuable advice. Stop advising. In that way, if, as and when the government would be ready to start interviewing managers, we would not be conflicted. We would now have to change our business model and become an investor.

The change of business model would need a change in structure, people, and capital base. We would need to build a team, a pipeline and have capital ready to invest in any fund that we created. I immediately set to work to make this happen. We had already secured the seeds of capital and help for this change from a specialist investor in 2010, interested to back emerging managers but recognizing the time it takes for them to emerge. In late 2011, looking to scale up, I found myself back where our journey in energy efficiency had started, across a board room table in Hong Kong. This time, we were with a prominent private equity and infrastructure investment group. They recognized the opportunity and our finance and service-led, relatively technology-agnostic approach. Rather than invest in China at the time, they said, focus on Europe and North America, where business and legal structures were well established and where we could prove the model and secure returns. They were well aligned but we had little time to lose, given the movement in the UK. We would need to agree a deal and act.

In the early winter months of 2012, I travelled to Davos, Switzerland, to land the deal with our potential partners from Hong Kong, who were attending the World Economic Forum. While we were consultants to the World Economic Forum, that does not translate into an invitation to the event itself. So, I took my invitation to a side event organized with the Environmental Defense Fund, trudged through the snow, and entered the world of Davos through the side door. And there, in the middle of town, between meetings with the Japanese Prime Minister, I met our potential partner from Hong Kong, and we shook hands on a deal.

Later in 2012, a week after my first son, Zachary, was born, we launched the first institutional energy efficiency fund in the UK with government backing. What became the Green Investment Bank, then still a division of the Department for Business, Innovation and Skills, committed £50 million to an energy efficiency fund, which would be unlocked if we could find another £50 million in matched funding. We went back to everyone that had said 'we will if you will'. Many of them did and, some 18 months later, we were fully up and running. Between the time that Zachary was born in 2012 and my second son, Charlie, was born in 2015, we experienced one of the most formative periods for the company. The model that we had established was working. Other governments in Singapore and Ireland and New York worked with us to replicate the model and establish funds. £100 million in 2014 in the UK became well over £200 million internationally by the time Charlie arrived a year later. We attracted capital from governments, insurance companies, pension funds, corporate strategic investors and private investors from all over the world.

We completed building the projects that we had invested in with our first government-backed fund in 2018. We were ready to exit and realize returns for investors. The path we chose was via an initial public offering (IPO) on the London Stock Exchange. While this had never been done before for energy efficiency infrastructure, I had designed and led the first such IPO for the wider infrastructure project asset class in 2006, a year before I left HSBC. The idea was that we could secure fair market value for our initial investors in the fund we were exiting, while at the same time creating a scalable platform for growth for new investors. The same bankers that supported our bid for the Green Investment Bank got behind the initiative and we went to market. December 2018 turned out to be the worst December on the London Stock Exchange since 1930. So success was sweet when we pulled it off.

Since 2018, we have grown the new company from the modest £100 million value at IPO to an equity market capitalization of over £1 billion and a constituent of the FTSE 250 Index by the summer of 2022. We have done so by raising additional equity capital around twice a year and investing it in projects. We have invested in LED lighting projects in commercial buildings in the UK and the USA that reduce energy demand by up to 90 per cent. We have invested in projects involving changing the heating, ventilation and air conditioning systems, building management systems and controls in hospitals and commercial buildings, in variable speed drives for warehouses and agriculture and in innovative cooling solutions for datacentres. We have invested in cleaner and more efficient energy supply by bringing energy directly to the point of use. We have done so with solar and storage on rooftops and carports and on land next to buildings. We have done so with gas engines and turbines, where we use the heat as well as the power produced, and where we source the gas not just from the natural gas grid but also from recycled waste gases, such as from the blast furnaces of steel mills. We have invested in renewable heat projects, from geothermal to sustainably sourced biomass. And we have invested in distribution networks, including major district energy projects in the USA and even Stockholm's gas grid, which now runs 80 per cent from biogas. We have also invested in electric vehicle charging infrastructure and in the green transport fuel sector in the USA and Europe through projects that generate renewable natural gas.

We are, however, just getting started. Complementing the London Stock Exchange listed investment company that we manage, we have launched a private equity infrastructure fund to invest at the development and construction phase of new projects. We have also launched a special purpose acquisition company, listed on the New York Stock Exchange, to help take a profitable private company, or a division of an existing industrial or utility group, public. We called it 'SDCL EDGE'. When I saw the name on the screen on the day we rang the bell on the floor of the New York Stock Exchange, it was a moving milestone on our journey. By October 2021 when it listed, we had come on a long journey from a standing start, from zero to having over US$2 billion of investment capital in the energy efficiency market.

But the next stage of our journey was just beginning. Rising prices, a massive push for decarbonization by companies and governments and increasing concerns around grid resilience had all been driving interest in energy efficiency. The lid was blown off each of those drivers by Russia's invasion of Ukraine. Soaring prices, setbacks for decarbonization and an energy security crisis to change them all pushed Europe and the world over the edge.

Notes

Introduction

1 *Nature* (2009) 'A safe operating space for humanity', https://www.nature.com/articles/461472a
2 Atul Arya, 'Whither energy transition?', Blog, 13 December 2019
3 Hesse, H. (1943) *The Glass Bead Game* (Holt, Rinehart and Winston)

Chapter 1

1 Widely reported. See, for example, Chivers, C.J. (2007) 'Russians plant flag on the Arctic seabed', *The New York Times*, https://www.nytimes.com/2007/08/03/world/europe/03arctic.html
2 See 'Opening remarks' by NATO Secretary-General Jens Stoltenberg at the start of the signing ceremony of the Accession Protocols with Finland and Sweden (2022) NATO, https://www.nato.int/cps/en/natohq/opinions_197759.htm?selectedLocale=en
3 There is even disagreement over spelling. Efforts have been made by Ukraine to persuade English-language media and organizations to exclusively use Kyiv (derived from the Ukrainian-language name Київ) instead of Kiev (derived from the Russian-language name Киев) as the name of the Ukrainian capital.
4 See ENGIE (2017) 'Nord Stream 2 AG and European energy companies sign financing agreements', https://www.engie.com/en/journalists/press-releases/nord-stream-2; Interfax (2022) 'Shell writes off $3.9 bln in losses from Russia exit; $1.6 bln from Sakhalin-2 and $1.1 bln from Nord Stream 2', https://interfax.com/newsroom/top-stories/78922/; Reuters (2022) 'Engie to book entire $1 bln credit loss from halted Nord Stream 2 pipeline', https://www.reuters.com/business/energy/engie-book-entire-1-bln-credit-loss-halted-nord-stream-2-pipeline-2022-04-21/; Reuters (2022) 'Germany's Uniper joins peers in writing down loan to Nord Stream 2', https://www.reuters.com/markets/europe/germanys-uniper-record-full-impairment-loss-nord-stream-2-loan-2022-03-07/

5 *Hurriyet Daily News* (2022) 'East Med gas feasible only through Turkey: Erdogan', https://www.hurriyetdailynews.com/east-med-gas-feasible-only-through-turkey-erdogan-170861
6 https://gr.usembassy.gov/statement-on-east-med-energy-cooperation/#:~:text=U.S.%20Embassy%20Spokesperson&text=We%20are%20shifting%20our%20focus,gas%20and%20renewable%20energy%20sources
7 See https://www.iea.org/reports/russian-supplies-to-global-energy-markets/oil-market-and-russian-supply-2
8 https://www.nato.int/cps/en/natohq/news_184633.htm
9 https://www.nbcnews.com/id/wbna19952934
10 *The Moscow Times* (2021) 'Russia warns West against Arctic encroachment ahead of talks', https://www.themoscowtimes.com/2021/05/17/russia-warns-west-against-arctic-encroachment-ahead-of-talks-a73924
11 *The Economist* (2022) 'How war in Ukraine is changing the Arctic', https://www.economist.com/the-economist-explains/2022/07/04/how-war-in-ukraine-is-changing-the-arctic
12 https://www.thearcticinstitute.org/increasing-security-focus-china-arctic-policy/
13 https://www.eceee.org/all-news/news/von-der-leyen-defies-russias-gas-blackmail-vows-eu-unity-with-poland-and-bulgaria/
14 DW (2022) 'Russia using gas as "a weapon against Germany"', https://www.dw.com/en/habeck-russia-using-gas-as-a-weapon-against-germany/video-62235336
15 Reuters (2022) 'World is in its "first truly global energy crisis" – IEA's Birol', https://www.reuters.com/markets/commodities/global-lng-markets-further-tighten-next-year-ieas-birol-2022-10-25/
16 *The Economist* (2022) 'Europe's winter of discontent', https://www.economist.com/leaders/2022/07/14/europes-winter-of-discontent
17 Gourntis, K. (2022) 'Time to separate the E from the S and the G', *Infrastructure Investor*, https://www.infrastructureinvestor.com/time-to-separate-the-e-from-the-s-and-the-g/
18 Energy Wire (2022) 'How Manchin-Sumer would change energy, from oil to solar', https://www.eenews.net/articles/how-manchin-schumer-would-change-energy-from-oil-to-solar/
19 RNZ (2022) 'Saudi Arabia: Biden raised Khashoggi murder with crown prince', https://www.rnz.co.nz/news/world/471056/saudi-arabia-biden-raised-khashoggi-murder-with-crown-prince
20 See https://www.worldstopexports.com/worlds-top-oil-exports-country/

Chapter 2

1 Burton, M. and Farchy, J. (2022) 'Why the nickel meltdown on the LME still matters: quick take', https://www.bloomberg.com/news/articles/2022-03-18/

behind-the-nickel-mess-on-the-london-metal-exchange-quicktake?leadSource=uverify%20wall

2 *Daily Post* (2022) 'Pelosi's Taiwan trip proves US impunity, reflects Ukrainian situation – Russia Foreign Minister Lavrov', https://dailypost.ng/2022/08/03/pelosis-taiwan-trip-proves-us-impunity-reflects-ukrainian-situation-russia-foreign-minister-lavrov/

3 See Reuters (2022) 'Russia's Potanin dodges politics and sanctions to flourish', https://www.reuters.com/business/russias-potanin-dodges-politics-sanctions-flourish-2022-05-04/; TASS (2022) 'Press review: Potanin blacklisted, but Nornickel spared and Erdogan rival gets jail time', https://tass.com/pressreview/1551383; Parry, A. (2018) 'Rusal sanctions: market turmoil and legal fall-out', Bulletin, 20 Essex Street, https://twentyessex.com/wp-content/uploads/2019/06/Rusal-sanctions.pdf; Spiegelberger, W.R. (2019) 'Anatomy of a muddle: U.S. sanctions against Rusal and Oleg Deripaska', Foreign Policy Research Institute, https://www.fpri.org/article/2019/04/anatomy-of-a-muddle-u-s-sanctions-against-rusal-and-oleg-deripaska/

4 IEA (n.d.) 'Mineral requirements for clean energy transitions', https://www.iea.org/reports/the-role-of-critical-minerals-in-clean-energy-transitions/mineral-requirements-for-clean-energy-transitions

5 European Commission Newsroom (2021) 'EU joins PDAC 2021 convention', https://ec.europa.eu/newsroom/growth/items/703201/en

6 World Bank Group (2017) 'The growing role of minerals and metals for a low carbon future', https://documents1.worldbank.org/curated/en/207371500386458722/pdf/117581-WP-P159838-PUBLIC-ClimateSmartMiningJuly.pdf

7 IEA (2022) 'Securing clean energy technology supply chains', https://iea.blob.core.windows.net/assets/0fe16228-521a-43d9-8da6-bbf08cc9f2b4/SecuringCleanEnergyTechnologySupplyChains.pdf

8 Yergin, D. (2020) *The New Map: Energy, Climate, and the Clash of Nations* (Penguin Press)

9 https://www.bbc.co.uk/news/world-asia-61768875

10 See https://climate-diplomacy.org/case-studies/civil-war-darfur-sudan

11 The World Bank (2022) 'Joint Statement: The heads of the World Bank Group, IMF, SFP, and WTO call for urgent coordinated action on food security', https://www.worldbank.org/en/news/statement/2022/04/13/joint-statement-the-heads-of-the-world-bank-group-imf-wfp-and-wto-call-for-urgent-coordinated-action-on-food-security

12 Ball, T. (2022) 'Black Sea blockade in Russia-Ukraine war threatens world food supply', *The Times*, https://www.thetimes.co.uk/article/black-sea-blockade-in-russia-ukraine-war-threatens-world-food-supply-k59bgxgjz

13 Panerali, K. (2021) 'Urban energy and the path to net zero', Presentation, 1 September

14 See United Nations (2013) 'UN report: one-third of world's food wasted annually at great economic, environmental cost', https://news.un.org/en/story/2013/

09/448652; United Nations (n.d.) 'Stop food loss and waste, for the people, for the planet', https://www.un.org/en/observances/end-food-waste-day

15 Ellen MacArthur Foundation (n.d.) 'The New Plastics economy: rethinking the future of plastics', https://emf.thirdlight.com/file/24/_A-BkCs_skP18I_Am1g_JWxFrX/The%20New%20Plastics%20Economy%3A%20Rethinking%20the%20future%20of%20plastics.pdf

16 Ellen MacArthur Foundation (n.d.) 'Rethinking business models for a thriving fashion industry', https://ellenmacarthurfoundation.org/fashion-business-models/overview

17 Sutherlin, M. (2022) 'Where does recycling actually go?', https://www.bloomberg.com/news/newsletters/2022-03-29/big-take-where-does-recycling-actually-go-how-tesco-plastic-program-works

18 In April 2022, Mathis Wackernagel and Leo Wambersie (research associate) wrote a paper discussing the implications of Russia's invasion of Ukraine for resource security. The paper used 'environmentally extended multiregional input-output' assessments to track the destination of exports to establish where crops and other products from Ukraine are finally consumed. The results are revealing. Israel, Jordan and Egypt all sourced over 20 per cent of total wheat and cereals from Ukraine. Tunisia, the country where the Arab Spring started in 2011, depended on Ukraine for nearly a quarter.

19 See https://data.footprintnetwork.org/?_ga=2.190504552.231308733.1657435268-602881615.1657085064#/

Chapter 3

1 Diamond, J. (2004) *Collapse: How Societies Choose to Fail or Succeed* (Penguin Random House)

2 See Zhang, Y. (2007) '"Warming of the climate system is unequivocal": highlights of the fourth IPCC assessment report', UN Chronicle, from Vol. XLIV, No. 2, 'Green Our World!', June 2007, https://www.un.org/en/chronicle/article/warming-climate-system-unequivocal-highlights-fourth-ipcc-assessment-report; IPCC (2007) Climate Change 2007 – The Physical Science Basis', Contribution of Working Group I to the Fourth Assessment Report of the IPCC, https://www.ipcc.ch/site/assets/uploads/2018/05/ar4_wg1_full_report-1.pdf; IPCC (2007) 'Climate Change 2007: Synthesis Report', Contribution of Working Groups I, II and III to the Fourth Assessment Report of the Intergovernmental Panel on Climate Change [Core Writing Team, Pachauri, R.K. and Reisinger, A. (eds.)], IPCC, Geneva, Switzerland, https://www.ipcc.ch/site/assets/uploads/2018/02/ar4_syr_full_report.pdf

3 The Nobel Prize 2007, https://www.nobelprize.org/prizes/peace/2007/summary/

4 Vince, G. (2022) 'Where we'll end up living as the planet burns', *Time*, https://time.com/6209432/climate-change-where-we-will-live/; Figueres, C. and Rivett-Carnac, T. (2020) 'What the world will look like in 2050 if we don't cut carbon emissions in half', *Time*, https://time.com/5824295/climate-change-future-possibilities/

5 World Meteorological Organization (2020) '2020 was one of the three warmest years on record', https://public.wmo.int/en/media/press-release/2020-was-one-of-three-warmest-years-record
6 *The New York Times* (2022) 'As the Great Salt Lake dries up, Utah faces an "environmental nuclear bomb"', https://www.nytimes.com/2022/06/07/climate/salt-lake-city-climate-disaster.html
7 Hodgson, C. (2023) 'Climate graphic of the week: Asia's prolonged April heatwave concerns scientists', *Financial Times*, https://www.ft.com/content/b1b07514-f1c0-4f1b-88ce-974379ce4d64
8 CleanTechnica (2021) 'Net zero emissions would stabilize climate quickly says UK scientist', https://cleantechnica.com/2021/01/04/net-zero-emissions-stabilize-climate-quickly-uk-scientist/
9 See https://www.mcc-berlin.net/en/research/co2-budget.html
10 See New Energy Outlook 2022, Bloomberg NEF, https://about.bnef.com/new-energy-outlook/
11 IPCC (2022) 'Summary for policymakers' (Cambridge University Press), Section C.3.2, https://www.ipcc.ch/report/ar6/wg3/downloads/report/IPCC_AR6_WGIII_FullReport.pdf
12 IPCC (2022) 'Summary for policymakers' (Cambridge University Press), page 300, https://www.ipcc.ch/report/ar6/wg3/downloads/report/IPCC_AR6_WGIII_FullReport.pdf
13 Adam, P.K., Amuakwa-Mensah, F. and Akorli, C.D. (2023) 'Energy efficiency as a sustainability concern in Africa and financial development: how much bias is involved?', *Energy Economics*, vol. 120, Elsevier, https://www.sciencedirect.com/science/article/abs/pii/S0140988323000750; World Resources Institute (n.d.) 'Accelerating building efficiency: eight actions for urban leaders', https://publications.wri.org/buildingefficiency/
14 Cembalest, M. (2023) *Eye on the Market Annual Energy Paper*, 13th Edition, JP Morgan
15 ISO New England's Future Grid Reliability Study, 29 July 2022, https://publications.wri.org/buildingefficiency/
16 https://www.eia.gov/outlooks/ieo/consumption/sub-topic-03.php
17 Hockenos, P. (2021) 'Germany's high-risk clean-energy balancing act', Energy Transition, https://energytransition.org/2021/05/germanys-high-risk-clean-energy-balancing-act/

Chapter 4

1 They were sued by ClientEarth. See Boffey, D. (2021) 'Court orders Royal Dutch Shell to cut carbon emissions by 45% by 2030', *The Guardian*, https://www.theguardian.com/business/2021/may/26/court-orders-royal-

dutch-shell-to-cut-carbon-emissions-by-45-by-2030; Gayle, D. (2022) 'Shell directors sued for "failing to prepare company for net zero"', *The Guardian*, https://www.theguardian.com/business/2022/mar/15/shell-directors-sued-net-zero-clientearth

2 Economy, E.C. (2006) C.V. Starr Senior Fellow and Director, Asia Studies Council on Foreign Relations, China's Environmental Challenge, Testimony Before the U.S.-China Economic and Security Review Commission, Hearing on Major Challenges Facing the Chinese Leadership

3 World Bank (2007) 'Cost of pollution in China', Washington DC, World Bank

4 *Financial Times* (2022) 'Asia-Pacific climate leaders 2022: interactive listing', https://www.ft.com/climate-leaders-asia-pacific-2022

5 IPCC (2022) 'Summary for policymakers' (Cambridge University Press), https://www.ipcc.ch/report/ar6/wg3/downloads/report/IPCC_AR6_WGIII_FullReport.pdf

6 IMF Blog (2022) 'Further delaying climate policies will hurt economic growth', https://www.imf.org/en/Blogs/Articles/2022/10/05/further-delaying-climate-policies-will-hurt-economic-growth

7 https://www.withouthotair.com/c19/page_114.shtml

Chapter 5

1 Voltaire (1770) *Questions sur l'Encyclopédie, par des Amateurs*. Vol. 2. Geneva, Switzerland, p. 250.

2 https://inters.org/files/einstein1905_photoeff.pdf

3 https://www.sciencedirect.com/science/article/pii/S254243511830446X

4 https://iea.blob.core.windows.net/assets/38ceb32d-9d49-4473-84c7-6ba803f8de08/NorthwestEuropeanHydrogenMonitor.pdf

5 https://about.bnef.com/blog/liebreich-the-next-half-trillion-dollar-market-electrification-of-heat/

6 Masters, B. (2022) 'BlackRock to launch energy security and transition infrastructure programme', *Financial Times*, https://www.ft.com/content/bfb1def6-fcba-4eab-8bf3-bb3a1624af40

7 *McKinsey Quarterly* (2022) 'The net-zero transition in the wake of the war in Ukraine: a detour, a derailment, or a different path?', https://www.mckinsey.com/capabilities/sustainability/our-insights/the-net-zero-transition-in-the-wake-of-the-war-in-ukraine-a-detour-a-derailment-or-a-different-path

8 Friedman, T. (2022) 'The Ukraine war still holds surprises. The biggest may be for Putin', *The New York Times*, https://www.nytimes.com/2022/06/07/opinion/ukraine-putin.html

Chapter 6

1 Buffett, W. (2013) The Essays of Warren Buffett: Lessons for Corporate America, Carolina Academic Press.

2 https://pubs.acs.org/doi/10.1021/acs.est.1c06458

3 https://flowcharts.llnl.gov/sites/flowcharts/files/2022-04/Energy_2021_United-States_0.png

4 Presentation to Decarbonising Cities Conference in Bern, Switzerland, by Kristen Panerali of the World Economic Forum, 1 September 2021

5 https://www.tesla.com/ns_videos/Tesla-Master-Plan-Part-3.pdf

6 https://iea.blob.core.windows.net/assets/0bb45525-277f-4c9c-8d0c-9c0cb5e7d525/The_Future_of_Cooling.pdf

7 https://publications.parliament.uk/pa/cm201719/cmselect/cmbeis/1730/173003.htm

8 Fraunhofer ISI (2015) 'How energy efficiency cuts costs for a 1-degree future', https://newclimate.org/sites/default/files/2015/11/report_how-energy-efficiency-cuts-costs-for-a-2-degree-future.pdf

9 European Climate Foundation (2022) 'New survey shows high support in Europe for energy efficient homes', https://europeanclimate.org/resources/new-survey-shows-high-support-in-europe-for-energy-efficient-homes/

10 McKinsey & Company, https://www.mckinsey.com/capabilities/sustainability/our-insights/the-net-zero-transition-in-the-wake-of-the-war-in-ukraine-a-detour-a-derailment-or-a-different-path

11 Euractiv (2022) 'EU Parliament groups unite behind 14.5% energy savings goal for 2030', https://www.euractiv.com/section/energy/news/eu-parliament-groups-unite-behind-14-5-energy-savings-goal-for-2030/; Euractive (2022) 'IEA presents energy efficiency push to make Russia's gas, oil obsolete', https://www.euractiv.com/section/energy/news/iea-presents-energy-efficiency-push-to-make-russias-gas-oil-obsolete/

12 IEA (2021) 'Net zero by 2050: a roadmap for the global energy sector', https://iea.blob.core.windows.net/assets/deebef5d-0c34-4539-9d0c-10b13d840027/NetZeroby2050-ARoadmapfortheGlobalEnergySector_CORR.pdf

13 IEA (2021) 'How energy efficiency will power net zero climate goals', https://www.iea.org/commentaries/how-energy-efficiency-will-power-net-zero-climate-goals

14 See https://www.iea.org/data-and-statistics/charts/co2-emissions-reductions-by-measure-in-the-sustainable-development-scenario-relative-to-the-stated-policies-scenario-2010-2050

15 IPCC Sixth Assessment Report, https://www.ipcc.ch/report/ar6/wg3/resources/spm-headline-statements/

16 Euractive (2021) 'IEA presents energy efficiency push to make Russia's gas, oil obsolete', https://www.euractiv.com/section/energy/news/iea-presents-energy-efficiency-push-to-make-russias-gas-oil-obsolete/

17 Phillis, A. (2022) EU nightmare. Putin handed victory as block admits "greed" thwarting plan to shun Russia', *Express*, https://www.express.co.uk/news/world/1623141/EU-news-chief-admits-greed-preventing-bloc-cutting-russian-energy-oil-gas-imports; Energy Monitor (2022) 'Energy efficiency equals energy security', https://www.energymonitor.ai/tech/energy-efficiency/energy-efficiency-equals-energy-security/

Chapter 7

1 https://www.presidency.ucsb.edu/documents/letter-all-state-governors-uniform-soil-conservation-law
2 Goldman Sachs (2021) 'Carbonomics', https://www.goldmansachs.com/insights/pages/gs-research/carbonomics-gs-net-zero-models/report.pdf
3 *The Economist* (2022) 'The Brazilian Amazon has been a net carbon emitter since 2016', https://www.economist.com/interactive/graphic-detail/2022/05/21/the-brazilian-amazon-has-been-a-net-carbon-emitter-since-2016
4 https://www.frontiersin.org/articles/10.3389/frsen.2022.825190/full
5 IPCC, https://report.ipcc.ch/ar6/wg2/IPCC_AR6_WGII_FullReport.pdf
6 Paul Hawken's 'Regeneration' and accompanying website, www.regeneration.org
7 https://www.bbc.co.uk/news/world-asia-india-61862035
8 https://essd.copernicus.org/articles/15/1675/2023/
9 Save on Energy (2022) 'American food production requires more energy than you'd think', https://www.saveonenergy.com/resources/food-production-requires-energy/
10 Speed, M. (2022) 'Ukraine war hits global timber trade and adds to risks for forests', *Financial Times*, https://www.ft.com/content/d6388b32-757b-4484-95ff-720b4b2319f3
11 The climate case, *Saul* v *RWE*, https://rwe.climatecase.org/en; Germanwatch, *Saul* v *RWE*, https://www.germanwatch.org/en/rwe; Source Material (2022) '"Battle of science" rages over Peru glacier', https://www.source-material.org/battle-of-science-rages-over-peru-glacier/
12 *Australian Conservation Foundation Incorporated* v *Woodside Energy Ltd & Anor*, http://climatecasechart.com/non-us-case/australian-conservation-foundation-incorporated-v-woodside-energy-ltd-anor/; ABC News (2022) 'Conservation group seeks injunction to stop Woodside gas project to protect Great Barrier Reef', https://www.abc.net.au/news/2022-06-22/scarborough-gas-project-faces-court-challenge-barrier-reef-fears/101169974
13 Environment Agency (2022) 'Water and sewerage company performance on pollution hits new low', https://www.gov.uk/government/news/water-and-sewerage-company-performance-on-pollution-hits-new-low
14 *The Economist* (2023) 'How to fix the global rice crisis', https://www.economist.com/leaders/2023/03/30/how-to-fix-the-global-rice-crisis

15 https://sustainability.emory.edu/wp-content/uploads/2018/02/InfoSheet-Energy26FoodProduction.pdf
16 Smil, V. (2022) *How the World Really Works* (Penguin)
17 Venkataraman, B. (2023) 'Country, the city version: farms in the sky gain new interest', *The New York Times*, https://www.nytimes.com/2008/07/15/science/15farm.html
18 https://www.science.org/doi/10.1126/science.aaq0216
19 See https://ourworldindata.org/food-waste-emissions
20 The World Bank (2019) 'Worsening water quality reducing economic growth by a third in some countries: World Bank', https://www.worldbank.org/en/news/press-release/2019/08/20/worsening-water-quality-reducing-economic-growth-by-a-third-in-some-countries; World Health Organization (2017) '2.1 billion people lack safe drinking water at home, more than twice as many lack safe sanitation', https://www.who.int/news/item/12-07-2017-2-1-billion-people-lack-safe-drinking-water-at-home-more-than-twice-as-many-lack-safe-sanitation; UN Sustainable Development Goals, 'Clean water and sanitation: why it matters', https://www.un.org/sustainabledevelopment/wp-content/uploads/2016/08/6_Why-it-Matters_Sanitation_2p.pdf
21 Terazono, E. and Hodgson, C. (2022) 'Food vs fuel: Ukraine war sharpens debate on use of crops for energy', *Financial Times*, https://www.ft.com/content/b424067e-f56b-4e49-ac34-5b3de07e7f08; The World Bank (2008) Policy Research Working Paper 4682, https://documents1.worldbank.org/curated/en/229961468140943023/pdf/WP4682.pdf
22 Gro Intelligence (2022) https://www.gro-intelligence.com/blog/gro-s-ceo-sara-menker-at-societe-generale-market-risk-and-the-agricultural-sector
23 Global Footprint Network (2022) 'Estimating the date of the earth's overshoot day 2022', https://www.overshootday.org/content/uploads/2022/06/Earth-Overshoot-Day-2022-Nowcast-Report.pdf
24 The Democrat (2008) 'Is our planet a global plantation?', https://www.natchez-democrat.com/2008/05/23/is-our-planet-a-global-plantation/

Chapter 8

1 Frankopan, P. (2023) *The Earth Transformed: An Untold History* (Bloomsbury)
2 IPCC Sixth Assessment Report, https://www.ipcc.ch/report/ar6/wg3/resources/spm-headline-statements/#:~:text=Electric%20vehicles%20powered%20by%20low,medium%20term%20(medium%20confidence).
3 Vivid Economics 2021, *Financial Times* (2022) https://www.atlanticcouncil.org/content-series/the-big-story/heat-is-killing-us-and-the-economy-too/
4 Lomborg, B. (2020) *False Alarm: How Climate Change Panic Costs Us Trillions, Hurts the Poor, and Fails to Fix the Planet* (Basic Books)
5 Wilson, T. (2022) 'Africa needs $25bn a year of investment to boost energy provision, says IEA chief', *Financial Times*, https://www.ft.com/content/6ee697a5-

fe5c-473c-9b0c-9b68dd200288; IEA (2021) 'Net zero by 2050: a roadmap for the global energy sector', https://iea.blob.core.windows.net/assets/deebef5d-0c34-4539-9d0c-10b13d840027/NetZeroby2050-ARoadmapfortheGlobalEnergySector_CORR.pdf

6 IPCC Sixth Assessment Report, https://www.ipcc.ch/report/ar6/wg2/resources/spm-headline-statements/

7 IPCC (2022) 'Summary for policymakers' (Cambridge University Press), https://www.ipcc.ch/report/ar6/wg3/downloads/report/IPCC_AR6_WGIII_FullReport.pdf

8 Kolwezi and Fungurume (2022) 'How the world depends on small cobalt miners', *The Economist*, 5 July

9 Quoted in: Kabir, H.M. (2010) 'Northern women development' [Nigeria]. ISBN 978-978-906-469-4. OCLC 890820657

Chapter 9

1 Telling, O. (2022) 'Malaysia's $30bn wealth fund to stand by carbon-emitting state companies', https://www.ft.com/content/549c3bee-adc2-451d-9488-e99b3314a60e; Natural Gas World (2023) '"Unleashing LNG" is the world's largest green initiative: EQT', https://www.naturalgasworld.com/unleashing-lng-is-the-worlds-largest-green-initiative-eqt-104498

2 Business Roundtable (2019) 'Promote an economy that serves all Americans', https://www.businessroundtable.org/business-roundtable-redefines-the-purpose-of-a-corporation-to-promote-an-economy-that-serves-all-americans

3 https://ballotpedia.org/Kentucky's_attorney_general_says_ESG_is_inconsistent_with_Kentucky_law_(2022)

4 https://www.flgov.com/2022/08/23/governor-ron-desantis-eliminates-esg-considerations-from-state-pension-investments/

5 Gelles, D. (2022) 'How Republicans are "weaponizing" public office against climate action', *The New York Times*, https://www.nytimes.com/2022/08/05/climate/republican-treasurers-climate-change.html

6 Walker, O. (2022) 'HSBC suspends banker over climate change comments', https://www.ft.com/content/8e1a16ea-bf63-45f8-81af-dc41c0df4e06

7 Moloney, C. (2022) 'HSBC Stuart Kirk suspended in climate row', *The Times*, https://www.thetimes.co.uk/article/hsbc-chief-suspended-in-climate-row-5s76fpqqb

8 See https://www.daines.senate.gov/wp-content/uploads/imo/media/doc/2022.06.09_Letter%20to%20HSBC%20re%20Stuart%20Kirk.pdf

9 GFANZ (2021) 'Amount of finance committed to achieving 1.5°C now at scale needed to deliver the transition', https://www.gfanzero.com/press/amount-of-finance-committed-to-achieving-1-5c-now-at-scale-needed-to-deliver-the-transition/

10 Climate Action 100, https://www.climateaction100.org

11 https://www.daines.senate.gov/wp-content/uploads/imo/media/doc/2022.06.09_Letter%20to%20HSBC%20re%20Stuart%20Kirk.pdf
12 https://www.ft.com/content/f1367ab4-ac6f-486d-8bd2-e7659448055d
13 Green News (2023) 'KLM axes "misleading" ads but won't stop promoting sustainability initiatives', https://www.euronews.com/green/2023/04/06/klm-airline-accused-of-greenwashing-adverts-which-encourage-responsible-flying-in-the-neth
14 ESG Global Study 2022, https://www.firstlinks.com.au/uploads/Whitepapers/2022/Capital-Group_esg-global-study-2022-chapter1-(en).pdf
15 Enterprising Investor (2022) 'ESG ratings: navigating through the haze', https://blogs.cfainstitute.org/investor/2021/08/10/esg-ratings-navigating-through-the-haze/
16 Dow Jones Sustainability World Index, https://www.spglobal.com/spdji/en/indices/esg/dow-jones-sustainability-world-index/#overview
17 https://www.eurosif.org/policies/sfdr/
18 Pucker, K.P. and King, A. (2022) 'ESG investing isn't designed to save the planet', *Harvard Business Review*, https://hbr.org/2022/08/esg-investing-isnt-designed-to-save-the-planet
19 ISS, European Sustainable Finance Survey (2020) https://sustainablefinance-survey.de/sites/sustainablefinancesurvey.de/files/documents/european_sustainable_finance_survey_2020_final_2.pdf
20 Van Steenis, H. (2022) 'Time to destigmatise "khaki finance"', *Financial Times*, https://www.ft.com/content/45b0cb96-9cb6-405a-b15c-99e599836fc0
21 Poll conducted by Uswitch, see https://www.capita.com/our-thinking/what-do-people-really-think-about-uks-ambition-become-net-zero#:~:text=Four%20in%20five%20say%20they,to%2042%25%20of%20female%20respondents
22 Survey conducted by the Local Government Chronicle, https://www.djs-research.co.uk/LocalGovernmentMarketResearchInsightsAndFindings/article/47percent-of-local-council-executives-confident-in-meeting-net-zero-targets-finds-survey-05129
23 Climate Change Committee (2022) 'Current programmes will not deliver net zero', https://www.theccc.org.uk/2022/06/29/current-programmes-will-not-deliver-net-zero/

Chapter 10

1 https://www.arup.com/perspectives/publications/research/section/drivers-of-change
2 Drivers of Change, Arup, 2007
3 IEA (2021) 'Net zero by 2050', https://www.iea.org/reports/net-zero-by-2050
4 IEA (2022) 'Securing clean energy technology supply chains, July, https://iea.blob.core.windows.net/assets/0fe16228-521a-43d9-8da6-bbf08cc9f2b4/SecuringCleanEnergyTechnologySupplyChains.pdf

5 https://blog.bizvibe.com/blog/energy-and-fuels/top-10-wind-turbine-manufacturers-world
6 https://www.nature.com/immersive/d41586-019-01124-7/index.html
7 https://journals.openedition.org/chinaperspectives/7216?lang=es
8 This is now the Cambridge Institute for Sustainability Leadership: https://www.cisl.cam.ac.uk
9 https://arxiv.org/pdf/1906.02243.pdf
10 https://www.energy.gov/eere/buildings/data-centers-and-servers
11 https://www.researchgate.net/publication/320225452_Total_Consumer_Power_Consumption_Forecast
12 Walls, A. (2022) 'Space rocket junk could have deadly consequences unless governments act', UBC News, https://news.ubc.ca/2022/07/12/space-rocket-junk-deadly/
13 https://www.cnbc.com/2022/02/11/elon-musk-spacexs-starship-is-solution-to-efficient-space-travel.html
14 https://now.northropgrumman.com/why-on-earth-should-we-be-mining-the-moon/
15 https://www.bbc.com/news/world-asia-china-61511546
16 Interview with *Financial Times*, July 2022, https://www.ft.com/content/94475cf3-df95-41bf-bef4-06e76a4df7df. The interview concluded that the 'UK places too little emphasis on reducing waste, through building efficiency, smaller cars and less food waste. Future technology won't save us; much of it already exists. What we need mostly is to use the technology'.
17 IPCC, Climate Change 2021: 'The physical science basis', https://www.ipcc.ch/report/sixth-assessment-report-working-group-i/
18 IPCC, Climate Change 2022: 'Impacts, adaptation and vulnerability', https://www.ipcc.ch/report/sixth-assessment-report-working-group-ii/
19 IPCC, Climate Change 2022: 'Mitigation of climate change', https://www.ipcc.ch/report/sixth-assessment-report-working-group-3/; See also the synthesis report at https://www.ipcc.ch/report/ar6/syr/resources/spm-headline-statements/

Chapter 11

1 Berkun, S. (2010) *The Myths of Innovation* (O'Reilly)
2 United Nations, Climate Change, https://www.un.org/en/global-issues/climate-change#:~:text=Climate%20Change%20is%20the%20defining,scope%20and%20unprecedented%20in%20scale
3 Reuters Institute for the Study of Journalism, Digital News Report 2020, https://reutersinstitute.politics.ox.ac.uk/sites/default/files/2020-06/DNR_2020_FINAL.pdf

4 Tyson, A. and Kennedy, B. (2020) 'Two-thirds of Americans think government should do more on climate', Pew Research Center, https://www.pewresearch.org/science/2020/06/23/two-thirds-of-americans-think-government-should-do-more-on-climate/

5 Tett, G. (2022) 'How do you persuade Republicans to save the planet?', *Financial Times*, https://www.ft.com/content/0becfebb-a15c-435f-86a5-64178884efb9

6 Tett, G. (2022) 'How do you persuade Republicans to save the planet?', *Financial Times*, https://www.ft.com/content/0becfebb-a15c-435f-86a5-64178884efb9

7 Pew Research Center (2021) https://www.pewresearch.org/short-reads/2021/07/23/on-climate-change-republicans-are-open-to-some-policy-approaches-even-as-they-assign-the-issue-low-priority/

8 https://www.bath.ac.uk/announcements/government-inaction-on-climate-change-linked-to-psychological-distress-in-young-people-new-study/

9 https://www.sciencedirect.com/science/article/pii/S0959378021001321

10 ResearchGate (2015) 'Climate change and the media', https://www.researchgate.net/profile/Mike-Schaefer-3/publication/304193905_Climate_Change_and_the_Media/links/5d0b44b4a6fdcc35c15bcdfd/Climate-Change-and-the-Media. pdf?origin=publication_detail

11 Memory and Cognition (2019) 'Science by social media: attitudes towards climate change are mediated by perceived social consensus', https://link.springer.com/content/pdf/10.3758/s13421-019-00948-y.pdf

Chapter 12

1 IPCC, Special Report: 'Global warming of 1.5 °C', https://www.ipcc.ch/sr15/

Bibliography

Numerous books, hundreds of articles and reports and thousands of citations, sources, experiences, and interviews were assembled and reviewed in the process of researching and writing this book, most of which are in the public domain. A selection of books is set out below.

Mike Berners-Lee, *There Is No Planet B* (Cambridge University Press, 2021)

Colin Butterfield and Jonnie Hughes, *Earthshot: How to Save Our Planet* (John Murray, 2021)

Mark Carney, *Value(s): Climate, Credit, Covid and How We Focus on What Matters* (William Collins, 2021)

Jared Diamond, *Collapse: How Societies Choose to Fail or Survive* (Penguin, 2006)

John Doerr, *Speed & Scale: A Global Action Plan for Solving Our Climate Crisis Now* (Penguin Business, 2021)

Peter Frankopan, *The Earth Transformed: An Untold History* (Bloomsbury, 2023)

Bill Gates, *How to Avoid a Climate Disaster: The Solutions We Have and the Breakthroughs We Need* (Penguin Random House, 2021)

Saul Griffiths, *Electrify: An Optimist's Playbook for Our Clean Energy Future* (The MIT Press, 2021)

Paul Hawken, *Regeneration: Ending the Climate Crisis in One Generation* (Penguin, 2021)

Dieter Helm, *Net Zero: How We Stop Causing Climate Change* (William Collins, 2021)

Bjorn Lomborg, *False Alarm: How Climate Change Panic Costs Us Trillions, Hurts the Poor, and Fails to Fix the Planet* (Basic Books, 2020)

Tim Marshall, *The Power of Geography: Ten Maps that Reveal the Future of Our World* (Elliot and Thompson, 2021)

Henry Sanderson, *Volt Rush: The Winners and Losers in the Race to Go Green* (Oneworld Publications, 2021)

Ernst Friedrich Schumacher, *Small Is Beautiful* (HarperCollins,1973)

Vaclav Smil, *How the World Really Works: A Scientist's Guide to Our Past, Present and Future* (Penguin Random House, 2022)

Daniel Yergin, *The New Map: Energy, Climate, and the Clash of Nations* (Penguin Random House, 2021)

Acknowledgements

Not every day of the last 15 years has been easy, but every day and every experience, good or bad, has contributed in one way or another to the contents of this book. I have had the great privilege to meet and spend time with an incredible number of great people, which has been a great inspiration. And I have been supported and believed in by great people, who have given me a great opportunity. I have been happy every day following my convictions.

But it is my wife, Laura, that has endured the last 15 years, and for that, and so many other things, I am forever grateful. She has not just supported me but survived everything with me. She is my roots and my strength and my love. And she has brought the two most wonderful children I could ever dream of, Zachary and Charlie, into this world, an incredible gift. We have been so lucky to have both of my parents, Martin and Julia, and both of her parents, Ann and Howard, behind our every step. My brother, David, has been on board from the earliest days of the journey. Laura's sisters, Natalie and Caroline, have travelled with us, and my brother-in-law, Ian, has been a beacon. Our nieces, Grace, Alice, and Elizabeth, have grown so much.

My business journey would simply not have been possible without the support of my partners and shareholders, Eli Shahmoon, Victor Chu, Elizabeth Kan, Gordon Power, Stephen Lansdown, Gareth Lake, Dai Ishida, Emanuel Citron, Cort Ahl, and Danny Stein. Tony Roper has been a guiding light chairing our London Stock Exchange listed investment company and Michael Naylor the catalyst for our New York Stock Exchange listed company, and so many other good things. Our independent directors and advisers, including Chris Knowles, Helen Clarkson, Emma Griffin, Sarika Patel, Steve Gilbert, Tony Davis, Robert Schultz, William Kriegel, Anna-Maria Fernandes,

Karl Olsoni, Tim Warrington, Andrew Lindsay and Guy Hands, have helped to shape and improve us. Michael Liebreich has been an excellent collaborator in constructive challenge and revolution from the inside. Our outside advisers, Richard Allsopp and Ronak Patel at Campbell Lutyens, have led our charge in private markets while Raphael Bejarano, Neil Winward, Tom Yeadon, Gaudi Le Roux and Tom Harris, together with their excellent teams at Jefferies, raised our first billion. Perhaps the hardest. Nigel Farr at our legal advisers, Herbert Smith Freehills, deserves my personal thanks for his support and partnership, as do Karl Heckenberg, Anish Mujmudar, Nadir Maruf, Fiona Daly, Fred Krupp, Jonathan Rose, Tony Malkin, Richard Branson, Werner von Guionneau, David Wright, Les Smith, Simon Conway, Gene Murtagh, Kenneth Matthews, Greg Barker, Nick Stern, Leslee Cowen, Andrew Ennis, Mary Canning, Masato Hisamune, Anthony Gutman, Ray Wood, Jason Scott and James Bevan.

Our team at SDCL has been a growing family and the firm has developed a culture that I am truly proud of. I only wanted to create the firm that I wanted to work in, and I am delighted to have succeeded, but it is all down to them. I fear mentioning one without offending all the others, but they are all on our website, and each of them knows my debt of gratitude. For bearing the load with me, I particularly thank Purvi Sapre, Eugene Kinghorn, Joseph Muthu, Sarah Miraj, Valerie Samuel, Ben Story and David Hourihane.

All the investors that have believed in and supported me and our funds over the years have provided us with an incredible opportunity, which we will never take for granted and always seek to return in thanks, in impact and in investment performance. I also thank all the clients for whom we provide clean energy services round the clock, round the world.

For their specific contribution to this book, I would like to thank Webster Stone of Rugged Land Media in New York, who challenged me to tell the story, and Ed Orlebar of TB Cardew, John O'Donnell of Rondo Energy and Joni Koch, Anjali Berdia and Francesca Lorenzini of SDCL, who challenged how I told it. I also thank my publisher, Jonathan Shipley, and Kizzy Thompson for supporting me and Jon Kirk, Anthony Harvison, Frances Prior-Reeves, Douglas Walker, Michael Smeeth and Michael Gould for encouraging me.

About the Author

Jonathan Maxwell is the Founder and CEO of Sustainable Development Capital LLP (SDCL), an independent investment firm with over US$2 billion of investment capital dedicated to efficient and decentralized generation of energy (EDGE). Jonathan has over 27 years' experience in business and finance with 15 years at the helm of one of the UK's only independent sustainability and climate-focused investment firms.

Jonathan, who maintains that 'if it's not commercial, it's not sustainable', led SDCL to establish first of a kind investment funds, often with government backing, in the UK, Europe, the USA and Asia. He has focused on energy efficiency, the largest and most cost-effective source of greenhouse gas emission reductions. He has advised global governments as well as attracted investment capital from a wide range of global investors. He pioneered with the first energy efficiency company of its kind on the London Stock Exchange in 2018, which is now a constituent of the FTSE 250 Index, as well as funds in the private capital market and a company listing on the New York Stock Exchange. SDCL was the winner of the Environmental Finance Boutique Investment Manager of the Year Award in 2021 and the SDCL Energy Efficiency Income Trust plc was winner of Citywire's Infrastructure Investment Trust of the Year in 2022. Also in 2022, SDCL and Jonathan were shortlisted as finalists for Investment Week's Sustainable Investment Awards: SDCL for Best Sustainable Specialist Fund and Jonathan for 'Outstanding Contribution to the Sustainable Investment Industry'. Jonathan is a member of the Reuters IMPACT Climate Council and co-leads the finance working group of the UK government's Energy Efficiency Taskforce.

Jonathan is a regular contributor to national and international press, radio, television, and online media, commenting on investment, financial markets and sustainability issues. Jonathan has a degree in Modern History from Oxford University. He lives in London, England, with his wife and two children.